The
original edition
of this book con-
tained oversized maps.
PDF files of these maps
can be downloaded from
www.elibron.com/maps

VALENTINE CHIROL

# TWIXT GREEK AND TURK

OR

JOTTING DURING A JOURNEY THROUGH
THESSALY, MACEDONIA, AND EPIRUS
IN THE AUTUMN 1880

Elibron Classics
www.elibron.com

Elibron Classics series.



ISBN 1-4021-5296-5 (paperback)
ISBN 1-4212-9511-3 (hardcover)

This Elibron Classics Replica Edition is an unabridged facsimile of the edition published in 1881 by William Blackwood and Sons, Edinburgh and London.

# 'TWIXT GREEK AND TURK

THE CASTLE OF YANINA.

# 'TWIXT GREEK AND TURK

OR

JOTTINGS DURING A JOURNEY THROUGH
THESSALY, MACEDONIA, AND EPIRUS,
IN THE AUTUMN OF 1880

BY

M. VALENTINE CHIROL

*WITH FRONTISPIECE AND MAP*

WILLIAM BLACKWOOD AND SONS
EDINBURGH AND LONDON
MDCCCLXXXI

# PREFACE.

---

The political events of the day have drawn so much of public attention to the regions which I have lately visited, that a sketch of their actual condition as I saw them in the autumn of last year, may not prove unacceptable at the present moment. I have not written in order to plead either this cause or that one. I am not a Philo-Turk, nor am I a Phil-Hellenic, or a Philo-Wallach, or a Philo-Albanian, or a Philo-Bulgarian; but I believe that a great change is at hand in the south-eastern peninsula of Europe, in which these rising nationalities have each its part to play. The old Ottoman empire—the Mussulman theocracy — is doomed, on this side of the Bosphorus at least. Whether its name and the shadow its power be still allowed to endure, or whether the "bag and baggage" policy be carried out to the bitter end, a new order of things is in process of evolution. If my jottings by the way can help the reader to a clearer apprehension of the grave events which must sooner or later remodel the map of South-eastern Europe, my journey will not have been undertaken in vain.

Some of the notes embodied in the following pages have already appeared in 'Blackwood's Magazine,' in the 'Pall Mall Gazette,' and in the 'Fortnightly Review;' and I am indebted to Mr J. Morley for his kind permission to make use of the last.

M. VALENTINE CHIROL.

CONSTANTINOPLE, *April* 8, 1881.

# CONTENTS.

NOTE.—The Sketch on the Cover of the book represents the remarkable aerial Monasteries of Meteora.

# 'TWIXT GREEK AND TURK.

## CHAPTER I.

### FROM CONSTANTINOPLE TO LARISSA.

It was in September last that I determined to carry into immediate execution my intention of visiting Thessaly, Macedonia, and Epirus. The first step to this end was naturally to obtain the permission of the Turkish authorities to travel in those provinces of the empire. And there my difficulties began. The naval demonstration was at its height, and the official world of Stamboul proportionately exasperated against foreigners in general, and Englishmen in particular, who were held to be singly and collectively responsible for this peculiar outcome of

Mr Gladstone's Turcophobia. In vain I petitioned for a *bouyourouldu* or passport—in vain I ante-chambered day after day in the crowded offices of the Sublime Porte, cheek-by-jowl with destitute widows and orphans of soldiers fallen in the late war for their Padishah, irate contractors soured by long-deferred payments, and all the host of beggars, creditors, and sycophants who every day hover with oriental patience about the tortuous corridors and gloomy chambers of the tumble-down building where the affairs of State have for so many centuries been managed or mismanaged. The *bouyourouldu* was not forthcoming. Excuses were, however, as plentiful as flies in June. Brigandage was rampant in those provinces, and the Turkish Government could not allow travellers to expose themselves to dangers such as Colonel Synge had but recently experienced, and from which he had been rescued only with difficulty and at great public loss. Moreover, the popular feeling in Thessaly and Epirus had been seriously excited by the recent action of the Powers, and a foreigner might find himself exposed to unpleasantnesses from which the local authorities could scarcely protect him. It was evident that the Porte did not wish to have busybodies roaming about the provinces and spying out the nakedness of the land. But unfortunately, I have ever been inclined to consult my own inclinations rather than my neighbours', and I certainly had no intention of deferring my plans to the convenience of the Sublime Porte. So I resolved to do without the vizierial *bouyourouldu;* and having equipped myself in the lightest marching order, and being provided with a few letters of introduction to local magnates, and to one of the commanding officers in Thessaly, I embarked on the 18th of September on board the Austrian steamer Apis for Volo, trusting to my own diplomacy and the chapter of accidents to enable me to develop my programme with better success at Larissa.

It takes as long to get from Constantinople to Larissa as to London. The Austrian Lloyd's and the French Fraissinet are the only two lines which connect the capital with the province of Thessaly; and their steamers spend six days, not altogether unpleasantly, in wending their way round the coasts of the Ægean to the Gulf of

Volo. The first morning after we lost sight of Seraglio Point, with its forest of cypress-trees and minarets, we halted at the Dardanelles, where General Blum's formidable new earthworks have usurped the name and function of Sultan Amurath's picturesque old fortress, the Key of the Seas. From the mouth of the Straits our course lay northwards between the mainland and the rugged islands of Imbros and of Samothraki. Here Neptune was wont of old to sit on the summit of Mount Saos, and survey the plain of Troy and the fleet of the Argives; and here the old sea-god doubtless woke up of late years to watch Britannia rule the waves in Besika Bay. All along the northern shores of the Ægean from Dédéagatch to Cavalla, between the sea and the wild mountain-range of the Rhodope, there stretches one of the richest tracts of land in European Turkey—the land of Yenidjé tobacco, for the choice crops of which there is every year the same keen competition in the market as for the *grands crûs* of Margaux and of Yquem. Both at Dédéagatch and Port Lagos we stop some hours to take in our fragrant cargo. But neither at the one nor at the other is any trace of the natural wealth of the district to be seen in the miserable, fever-stricken towns. A few warehouses and merchants' dwellings along the shore, with rows of squalid hovels, alone mark those geographical expressions which recent history has made famous by the disastrous winter retreat of Suleiman Pasha's army from Adrianople across the Rhodope. Cavalla alone escapes the curse of the malaria which infests these fertile plains; for it lies at the western limit of this tobacco region, where an abrupt spur of the Rhodope projects into the sea: a smart, clean little town, built up against the side of a precipitous rock—white houses interspersed with the dark-green of myrtle and orange groves, and the domes and minarets of churches and mosques rising in terraces above the sheer sea-cliff, under the shadow of a picturesque old fortress, which is certainly more adapted nowadays for ornament than for use: a wealthy town withal, where Greek traders and Turkish landowners reside in that good-fellowship which is the usual outcome of common interests and mutual usefulness, undisturbed by political intrigues and official solicitations. Happy, too, is Cavalla, insomuch that, save for its being the birthplace of Mehemet Ali, it has no history. But the memory of the great Egyptian ruler who once made the Padishah tremble on his throne still lives in Cavalla, and a spacious building with stately arcades and terraces marks the place where the first Khedive was born. Mehemet Ali—who, by the by, was of Albanian lineage—mindful in after-days of his native town, built and endowed this establishment, where, according to the founder's wishes, gratuitous instruction is still given to a certain number of Mussulman children, and where humble wayfarers of all creeds and races may obtain food and shelter during their sojourn in the town. As we steamed out of its peaceful little bay the shades of evening were slowly creeping over the quaint old town. But the warm sunlight still lingered on the neighbouring island of Thasos; and the long ridge of Mount Ipsario was flushed with gold, recalling the days when the Phœnicians worked its goldmines, and Archilochus described it, in allusion to its wealth and shape, and perhaps also to the intelligence of its inhabitants, who allowed foreigners to reap their treasures, as an ass's backbone cased in gold. As we drew out into the open, the whole range of the Rhodope came again into view, the soft after-glow resting peacefully on the slopes and vales, which but a few years ago were lit up by the lurid glare of burning towns and villages. In the whole history of the last Turco-Russian war, there is no more thrilling chapter than that which records the long and heroic resistance offered by the Moslem populations of the Rhodope to the advance of the victorious Russians. The Treaty of San Stefano had been signed, and the Rhodope was included in the area given up to the tender mercies of the Russian army of occupation. But its inhabitants knew too well what they had to expect at its hands: the thousands and tens of thousands of miserable refugees from Bulgaria and Roumelia who had fled at the Cossack's approach to its mountain fastnesses, had too often told the same harrowing tales of wanton outrage and burning homesteads. Abandoned to their fate by the rulers of Constantinople, without either the materials or the sinews of war, they resolved at least to die in defence of their homes. A few foreign enthusiasts came to give them their help in organising their resistance; and for five months this half-armed mob of peasantry kept the victorious armies of the Czar at bay. Now and then the Cossacks made a successful raid into their mountains, carrying fire and sword into the unfortunate villages; but the Russians, repulsed in one or two important engagements, failed to make good their foothold. Fearing to risk their laurels in a desperate, and, to them, barren struggle, and dreading the political consequences which it might have involved, they contented themselves with blockading the so-called insurgents in their mountain fastnesses; and the Rhodope, surrounded on all sides by Russians, remained untrodden by their soldiery, like a solitary island, against which the waves of invasion beat up in vain.

Midnight brought us up under the lee of the Holy Mountain, and the lofty cone of Mount Athos towered white and ghostly above us, rising abruptly from the sea to a height of over 6000 feet. In the morning we were fairly in the Gulf of Salonica; and to the west, on the distant shores of Thessaly, the Cloud-compeller looked down upon us from the dome-shaped summits of Olympus, where he sits in majestic exile awaiting the day when the Conference of Berlin and the concert of the Powers shall restore him to his chosen people.

A day at Salonica, the true Jerusalem on the sea—where 40,000 Hebrews, descendants of one of the many batches of unfortunate Jews expelled from Spain at the behest of the Grand Inquisitor, still assert by their Spanish idiom and Hebrew type the exclusive purity of the Semitic race—and at last, on the fifth morning after leaving Constantinople, the good steamship Apis rounds the jagged spur which Pelion projects like a huge claw across the mouth of the Gulf of Volo. Encircled by bleak and lofty mountains and steep cliffs, which only fall away towards the north-west to allow access to the plains of Thessaly, the gulf is sheltered like a lake from the winter storms which have made the Ægean *famosum tempestate mare;* though now and again impetuous squalls rush down the mountain gullies and lash its deep-blue waters into transient fury, to the discomfiture of the puny native craft. Presently the town of Volo itself comes into sight, stretching in a long white line round the head of a small creek-like bay at the most northern end of the gulf. A considerable Turkish squadron — two ironclads, a corvette, and a couple of despatch-boats — lie at anchor before it, watching sullenly over the town which is soon to be wrested from their grip. Perched up aloft against the barren mountain-side, some 1500 feet above the town, are the villages of Macrinitza and Portaria, famous for the desperate stand which they made against the Turkish troops during the short-lived insurrection of 1878. Since then the Turks have constructed earthworks on the crest of the hills above them, and they are virtually at the mercy of those batteries.

Volo, like many Turkish towns along the basin of the Mediterranean, is really composed of two distinct towns—the old Turkish and the modern Christian town. The old town of Volo lies north, at the head of the bay. Inside a ruinous fortress, where a few obsolete cannon still peep over the tumble-down walls, there is a Turkish bazaar, some barracks, and perhaps a hundred houses, almost entirely tenanted by

Mussulmans, to the number of 1200 or 1300. But I was not allowed to explore it; for the recent visit to Volo of the British military *attaché* had roused the suspicions of the garrison, and no sooner was my *giaour* head-gear detected inside the gates than I was courteously asked my business. On my replying that I wished to pay my respects to the *bimbashi* in command, I was told he was from home, and requested to call another day. So I was only able to take stock of a small wooden pop-gun, which appeared to have been placed inside the fortress-gate to enfilade inquisitive intruders. Across a marshy swamp, where a very thin battalion was encamped, presumably upon sanitary considerations, I trudged back to the new town of Volo. The latter is scarcely more than thirty years old. One of its veteran inhabitants, who settled there in 1858, told me that at that time it counted only about 80 houses; whereas now there are over 500 houses, with a population of some 4000 souls, Greeks almost to a man. It consists of one long, broad street, more than half a mile in length, running parallel to the sea, with a few warehouses and a steam-mill, and some shipping agencies, along the water's edge and in the streets running down to it. The houses are mostly well built, of stone and stucco, in the modern Greek style ; and, besides coffee-houses and wine-shops innumerable, it boasts even a hotel, whereof the less said the better. Annexation to Greece is naturally looked forward to here with unmixed satisfaction. No town in Thessaly would gain more by the change. The only outlet for the produce of this rich grain-growing province, with an import and export trade which already exceeds £350,000 per annum, with a shipping traffic of nearly 150,000 tons, endowed by nature with an admirable roadstead and easy means of developing communication with the interior, it would soon grow into a first-class port, and take tithes of the increased prosperity which must accrue to the whole province. But there is an entire absence of anything like political excitement. The cession of Volo to the Hellenic kingdom, already accepted by the Ottoman Government in its former proposals, has long been looked upon as a foregone conclusion by Greeks and Turks alike. There is neither exultation on the one hand, nor despondency on the other, both having been long since discounted.

Between Volo and the capital of Thessaly there is a carriage-road, but, like most such roads in Turkey, heaven-made — the hand of man has had little to do with it; and he who allows himself to be jolted over it for eight mortal hours, in the rude wooden canopied carts called *brashowkas*—be the hay stretched under him ever so soft, his supply of rugs and wraps ever so plentiful—has cause to recollect for many a day the "Sultan's road" from Volo to Larissa. On reaching the crest of the hills which divide off the seaboard from the plain of Thessaly, one obtains a most lovely view of the town and Gulf of Volo. The bastions and minarets of the Turkish quarter; the white houses of the new town; the groves of olive-trees, nestling at the foot of the mountains; the numerous villages, perched like eagles' nests on the precipitous slopes of Pelion; the bold outline of the jagged coast; the smooth blue waters, studded with many a quaintly rigged sail, and the dark forms of the Turkish fleet riding at anchor in the roadstead,—make up a striking picture, full of lights and shades, both for the moralist and the painter. Down the other side, the road soon reaches the vast treeless plain of Thessaly, which stretches away from the foot of Pelion and Ossa in one unbroken reach to the distant heights of Trikalla. Past a few miserable villages, and scarcely more prosperous *chiftliks*, through field after field of stubble, and wide tracts of land lying fallow, with here and there a marshy swamp, the *brash-owka* jolts along for six weary hours towards the distant minarets which mark Larissa, and seem to recede phantom-like as we advance. But to all misery there is a term. By the time one has stuck his cramped extremities twenty times, first out of one, then out of the other, side of the strangely devised vehicle, in which there is room neither to sit nor to recline full length, but only to lie doubled up like a trussed fowl, our three horses put on a sudden spurt, and the giddy chariot bowls jauntily through the mud-gateway pierced in the mud-embankment which forms the entrance to Larissa, and threads its way amid mud-houses and mud-hovels through the tortuous streets of the capital of Thessaly.

# CHAPTER II.

## THESSALY AND ITS CAPITAL.

NOR is there much to repay the weariness of the journey when the capital of Thessaly is at last reached. With its mud-rampart, mud-walls, mud-houses, and a few whitewashed buildings, barracks, *conaks,* and mosques rising out of a yellow, treeless plain, it resembles nothing more than an Arab town of Upper Egypt shorn of its palm-trees and of the magic of its climate. To make the resemblance still more striking, there were groups of negroes squatting about the road just outside the gate through which we entered; for it was Friday afternoon, and the black population of Larissa had turned out in full festival-attire to bask in the autumnal sunshine—men, women, and children in gaudy Manchester cottons, contrasting strangely with their swarthy complexions. Larissa swarms with negroes, children and grandchildren of liberated slaves: many also have been brought here as slaves and have purchased their liberty or claimed it of late years from the authorities and foreign consuls in virtue of slave-trade treaties and recent legislation; while not a few are still in bondage. But it is only fair to say that Larissa is not viewed at its best from either the Volo or the Salonica roads. As in other and fairer cities, the fashionable quarter of Larissa is the west end. There the Salemvria rolls its lazy, yellowish-green waters under a Byzantine bridge of five handsome spans; a few trees grow along its banks; and above them, on a solitary hillock, a picturesque old mosque, bedomed and beminareted, rising almost side by side with the Greek cathedral and the palace of the Despot or Archbishop of Larissa, two or three Government buildings, and a quaint, square clock-tower of Latin construction, form an imposing group, over which looms in the far distance the clean-cut angular peak of Ossa. Should the traveller wish to carry away with him a favourable impression of Larissa, let him halt here when the last flush of sunset still lingers over the scene, and not court disenchantment by passing across the threshold of the town. Crooked, straggling streets, ill paved, and studded with yawning pits and pools of stagnant water; rows of mud-walls; dirty, rickety bazaars, stocked with the refuse of the Manchester and Vienna markets,—form an *ensemble* than which none more dreary can well be found in the whole length and breadth of the Ottoman empire. Verily, if its name be correctly interpreted to mean "The Brilliant," what a falling off is here since the days when it deserved its title!

Yet Larissa is the capital of one of the richest provinces of the empire; it counts between 20,000 and 25,000 souls; it is the residence of high civil and military functionaries; it is the central *entrepôt* of the grain trade of the Thessalian plains; among its inhabitants are to be found some of the wealthiest landowners in Turkey; and even in the humblest homesteads there is a degree of ease and comfort which would raise envy in many another country. But the incubus of Turkish misrule has paralysed all wholesome activity and checked all desire for improvement, and under the present circumstances the Turkish authorities have other things to think of besides reforms and public works. Every nerve has to be strained to prepare for the threatened campaign. Though, at the time of my visit, the forces collected in Thessaly were not numerically imposing (indeed they were weaker than a few months before, as a large number of time-expired soldiers, whose grumbling, or worse than grumbling, at long years of unpaid service, threatened to spread disaffection among the troops, had been weeded out and sent back to their homes, and their places not yet filled), the first and second ban of the reserves have since been called out, and Turkey must now have a seasoned army of 40,000 men under arms in the province. But the real difficulties with which the Government has to grapple are of another and more stubborn order—viz., financial and administrative. Thessaly, as I before stated, is one of the richest provinces of the empire. In good years its crops of wheat, barley, Indian corn, and tobacco alone represent more than a million sterling. The tithes and muttons (as the tax on live stock is called) bring in 200,000 *liras*, and the local expenditure has been cut down within such narrow limits that the provincial revenue shows an annual surplus of 250,000 *liras*. Yet withal, the province is practically bankrupt. Constantinople devours all its wealth, and leaves to it only the burden of its debts. The sale of Crown property has been going on briskly since last summer. The principal purchaser has been Abram Pasha, the agent of the ex-Khedive at Constantinople, who is rapidly becoming one of the wealthiest landowners in Turkey. Large estates have also passed into the hands of native Greeks, as well as of Greeks from the Hellenic kingdom. Nearly half a million has been paid by the purchasers, but the provincial exchequer remains empty as heretofore, for the Porte swallows up every farthing of the purchase-money. To Government contractors alone the authorities at Larissa owed in October last 150,000 *liras*; and so absolutely had the uncertainty of the political situation and the unscrupulous financial measures of the present Grand Vizier destroyed the last vestige of local credit, that, instead of selling, as usual, the proceeds of the tithes (which are collected in kind), the Government was obliged to keep them in store in order to have wherewithal to feed the troops. Already maize was being served out to them in lieu of rice, because the contractors refused to renew the supplies on any terms save cash, and cash there was none to give.

In Larissa itself the Greeks are in a minority, the proportion being about 13,000 Moslems to 8000 Greeks and 3000 Jews. But in the whole district of Larissa—albeit that Larissa is, of all the districts of Thessaly, the one in which the Moslem or Turkish element is most strongly represented — the proportions are more than reversed. The following is a table of the populations of Thessaly compiled from the most trustworthy sources :—

| | Christians (Greeks). | Moslems (Turks). |
|---|---|---|
| Larissa, . . . . | 65,000 | 21,000 |
| Volo, . . . . | 51,000 | 4,000 |
| Trikalla, . . . | 70,000 | 3,500 |
| Karditza, . . . | 48,000 | 4,800 |
| Elassona, . . . | 32,000 | 5,500 |
| Pharsala, . . . | 17,000 | 4,000 |
| Armyro, . . . | 9,000 | 2,500 |
| | 292,000 | 45,300 |

To these 292,000 Christians and 45,300 Mussulmans must be added some 6000 Jews, thus forming for Thessaly a grand total of 343,300 souls.

The Christians of Thessaly, with the exception of about 20,000 Wallachs in the district of Elassona, as many in the district of Trikala, and a sprinkling of the same race in the northern villages of the district of Larissa, are all Greeks, and they exhibit most of the qualities as well as most of the defects of the Hellenic race. Especially in the towns, they are intelligent, industrious, pushing, and gifted with a rare commercial instinct. The whole commerce of Thessaly, which even now exceeds £600,000 per annum for imports and exports alone, apart from local trade, is in their hands. The few native industries of the province, such as cotton and woollen tissues, dyed stuffs and prints, shawls, and carpets, many of them showing fairly good taste and workmanship, are almost without exception Greek or Wallach. The schools, which have been founded and developed throughout Thessaly principally by patriotic donations from without, have already begun to rouse among the younger generations that innate love of learning which is one of the best features of the Hellenic race. At the present moment there are seventy-eight higher schools and sixty-one elementary schools in the province, where instruction—albeit of a very rudimentary nature—is given to upwards of 10,000 children. It is at least a commencement, and many parents among the wealthier classes are thereby induced to send their children to complete their education in Greece, whence they return as doctors, engineers, architects, merchants, schoolmasters, &c., to stimulate others by their example and successes. Why is it that the Greek peasant of the plains, in spite of his indisputable material prosperity, seems to be stricken with a sort of moral and mental paralysis? No doubt misgovernment weighs heavily upon him, but it does not suffice to account for the phenomenon. The clue is to be found in the fact that "the rich and fertile plains of Thessaly, which are almost without exception *chiftlik*—*i.e.*, freehold property of beys and others—were formerly village communes, confiscated and sold by Ali Pasha of Yanina at the end of the last century; and the peasants, who cultivate them now as tenants, are the descendants of the former proprietors, whose names have still adhered to many of their fields."—(Notes by Colonel Synge respecting the state of the Peasant Farmers in Thessaly. *Parliamentary Blue-book — Greece, No.* 1: 1880.) It is the memory of recent spoliation, not the oppression of the present landlords, which weighs down the peasant, which makes him treacherous, slothful, and vindictive, careless of small improvements, and impatient for a radical change. It is easy to denounce the despotism of Mussulman landowners. But religion has nothing to do with the question. At the present day nearly a third of the land in Thessaly is owned by Christians, and up till lately the Turkish beys were better liked as masters than the Christian landowners. Possessing generally enormous estates, from which they derived far larger revenues than they required for their expenditure, the chief items of which arose from the profuse hospitality they exercised, they were not inclined to infringe upon the share of the profits which accrues under the *métayer* system to the farmer. If the peasant was wronged or oppressed by Government officials, the influence which the beys enjoyed with the local authorities was generally exercised on his behalf to procure redress, and not without success. The beys, in fact, knew full well the value of the hen which laid them golden eggs, and took proportionate care of it. Of course, much depended upon the personal character of the bey; but as a rule, and within certain limits, he was an easy-going master. The Christian landowner, on the other hand, especially when an absentee, was harder to deal with. He had larger wants, and was therefore more grasping; he was more intelligent, and therefore more fond of pestering for improvements; he was more exposed to the extortions of the Government, and therefore more anxious to recoup himself upon the peasantry, while he was powerless to defend them against the abuses of those in authority. But of late years a great change has come over the attitude of the Turkish bey. Increased taxation, the constant demands of the Government for fresh loans, war contributions, and so-called voluntary gifts, the rapacity of officials intensified by long arrears of pay, have weighed heavily upon him, and his hand has in turn been heavier upon the peasantry. Then recent political changes, the assertion of the rights of the Christian populations of Turkey, and especially the Greek claim to Thessaly, have induced him gradually to look upon his Greek peasant as an impatient heir to his estates, and therefore as an enemy. Last year the cattle disease ravaged Thessaly; more than 30,000 head of oxen were carried off; and the peasants, already reduced by two successive bad seasons, were unable to replace the oxen, without which their fields must lie fallow. But the bey, who had hitherto been generally ready to advance money to his farmers on reasonable terms, hardens his heart against them, and declines to loosen his purse-strings. "Who knows," he says, "what is going to happen? You will repay me next year? But perhaps by next year we may all have been turned out of Europe: will you follow me then to Asia to repay my loan? My estates suffer? Let them suffer; who knows how long they may yet be mine?" There are, however, indications that this feeling is but a transient and pardonable ebullition of ill-temper. After the decisions of the Berlin Conference first became known, there was a good deal of tall talk about local resistance. But that soon died away. Then the beys declared that they would sell their lands and emigrate *en masse.* A few, very few, did so. Most of them thought better of it, knowing full well that their property could only gain in value by the change. Probably many of them will leave the country for a time should it actually pass into the possession of Greece, but they will retain their estates, and leave them in charge of Greek agents. This would not by any means satisfy the peasant. It is not the present state of the land laws which provokes his hostility. The *métayer* or partnership system, under which most of the estates are farmed, is a very simple one, and not unfavourable to the tenants. The proprietor of the estate furnishes the land, house, and seed to the tenant-farmer, and the latter is obliged to find the labour, oxen, and instruments. No written contract is entered into between them; and while the landlord has the right to discharge the tenant at his pleasure, the tenant is equally free to depart when he chooses. But evictions are very rare; and the landlord is forced, by the scarcity of labour and the absence of machinery to replace it, to make every possible concession in order not to lose his tenants. According to the *mêtayer* system, the produce of the farm is divided into two equal portions between the landlord and the tenant, after first deducting the seed furnished by the landlord and the tithe due to the Government. In some estates where the seed is furnished by the tenant, he even takes two-thirds of the produce, leaving only one for the landlord. No doubt the position of the peasantry is aggravated where the landlord lets out the whole property to a middleman, or *multazim*, who derives what revenue he can out of it by subletting it. But the *multazim's* exactions never impoverish him so much as the enforced idleness of the many feast-days prescribed by the Church. So numerous indeed have they become, that his working days—making no allowance for detention by weather or sickness—are limited to less than 200. "The agent of the largest proprietor in Thessaly," says

Colonel Synge, "informed me that he made a bargain with the labourers employed on his estates for 58 feast-days per annum, besides Sundays; and," he adds, "there is little doubt that these days of idleness, and frequently of drunkenness, are one of the great drawbacks to the prosperity of the country."

It is not the wrongs which he suffers at the hands of his landlords which make the Greek peasant yearn for union with the Hellenic kingdom. It is, that he sees in that consummation the prospect of the reopening of the whole question of land tenure in Thessaly. He treasures up not only the memory of the past, when his ancestors owned the land which he now tills as a hired labourer, but in many cases the very title-deeds of their property; and he looks forward to the day when Greek annexation shall restore life to the dead letter of the musty parchments, and the landlord shall be evicted in favour of the farmer. Though such hopes are undoubtedly doomed, under all circumstances, to disappointment, they have been often encouraged by unscrupulous agitators to stir up the popular passions. But the weapon is a double-edged one, and cuts both ways. The hopes which allure the farmer are construed by the landlord into a threat. The Mussulmans are naturally opposed to a change which would rob them, at least in a great measure, of their privileges and power; but the Christian proprietors would be equally opposed to a change which might rob them of their broad acres. It is not, however, the sympathies or inclinations of one or other section of the population which will decide the fate of Thessaly, as neither the one nor the other seems disposed to shape it by their own action. Supineness is the chief characteristic of Mussulmans as well as Christians. There is on both sides great eagerness to know what is going to be done for them or with them, but very little eagerness to do anything for themselves. The chief feeling, indeed, throughout the crisis has been—"Let things end as they may, but for God's sake let there be an end to the present intolerable uncertainty."

# CHAPTER III.

## UNDER THE SHADOW OF OLYMPUS.

NOTWITHSTANDING the kind hospitality which I enjoyed at Mr Longworth's, H.M.'s Vice-Consul, I had not been many days at Larissa before I felt the keenest anxiety to depart from it. I had been interviewed by the leading members of the various communities, who were anxious to let me have their views on the political situation of the province. The iniquity of Moslem rule had been vigorously denounced over coffee and cigarettes by patriotic Hellenes. Coffee and cigarettes had inspired the eloquence of Turkish landowners protesting against the revolutionary tendencies of Greek annexationists. I had heard the naval demonstration applauded and abused in turn, in terms quite as energetic as could be found in the columns of

Tory or Radical prints; I had dined with the Greek Consul; I had dined with the Governor; I had dined with the Archbishop; I had dined with the Commander-in-chief. The gaieties of Larissa began to pall upon me, and I determined to move on. The district which I was most anxious to visit was that which lies under the western slopes of Mount Olympus, and through which runs the only practicable road between Thessaly and western Macedonia. Nothing could exceed the courtesy and the alacrity displayed by the Turkish authorities to assist me in carrying out my plans. But all the mountain regions of Thessaly were infested with brigands. Greece had freed her own provinces from this plague at the expense of the neighbouring provinces of Turkey, where she conveniently discovered that the black sheep who were so unwelcome within her own borders, could be used with considerable effect to stir up the Greek populations and damage the little credit which Turkish rule might still enjoy in the eyes of Europe. To be captured by brigands is no doubt a novel and fascinating sensation: but excitement may be purchased too dearly, and I was not inclined to pay for it with the loss of my nose or my ears; and I doubted whether a paternal Government would again show itself as lavish, even of other people's money, as when it paid £12,000 for the release of Colonel Synge and invited the Turkish Government to refund the amount. On the other hand, I did not like to ask the authorities for a small army to accompany me on an excursion of mere curiosity. It was here that Providence favoured me with one of those fortunate opportunities by which it seeks intermittingly to atone for its more frequent unkindness. I had brought from Constantinople a letter of introduction to Selami Pasha, the general officer then in command of the cavalry division in Thessaly, and it was at that time borne in upon the Sublime Porte that he should be transferred to the same duties in Macedonia. His Excellency's route lay through the region which I wished to visit, and he at once invited me to accompany him to Monastir. I accepted the invitation.

Larissa, "The Brilliant," looked even more than usually squalid when I turned out of the

Consulate on the morning of the 30th of September to join the Pasha. Heavy rain had been falling throughout the night: it lay in the streets in large black flakes; it ran down the mud-walls of the mud-houses; it dripped in heavy drops from the low tiled roofs; it rushed down the open drains of the miserable bazaars in turbid torrents, sweeping before it the accumulated filth of weeks; it hung in lowering clouds over the town, shutting out the only redeeming feature of the landscape, the prospect of the classic mountains which surround the plain of Thessaly. The Pasha's orderly, who came round to fetch me, told me that his Excellency was ready, and waiting for me to start. But the word "ready" has in Turkish a most elastic sense, varying according to the intonation from the immediate present to a somewhat indefinite future. It was evidently the latter signification which attached to it in this case; for when I entered the courtyard of the Pasha's house, I found him sitting with a crowd of other officers on piles of baggage, waiting, not for me, but for his pack-animals. Compliments, cigarettes, and coffee helped, however, to while away the time. In the course of about an hour the animals put in an appearance. In the course of another hour we were, one after another, robbed of all our seats, and the caravan was finally pronounced ready in the fullest sense of the word. With a great show of haste and clatter of hoofs we trotted through the town, bespattering with mud the Greek merchants, who scowled at the military uniforms from behind their counters, and drawing to the windows and doorways respectable Jewish matrons, forgetful, in the hurry of the moment, of their matrimonial wigs, and dark-eyed Moslem ladies with tantalising veils, and sallow-faced Greek girls without. Presently we crossed the bridge over the Salemvria, and halted for a few minutes in the plain beyond to take leave of the chief officials and inspect the troops, who were drawn up in honour of the departing Pasha. Four battalions of infantry, two squadrons of cavalry, and four batteries of artillery had turned out for the occasion. The uniforms were patched and soiled, the horses small and ill-matched, but the men looked resolute and workmanlike enough. The troops, indeed, presented rather the appearance of having just returned from a campaign than that of an army preparing to encounter its hardships. The Pasha accomplished the review in a rather perfunctory manner. More coffee and cigarettes were consumed, and then the band struck up the Imperial hymn, and we cantered away across the plain. Our party was now considerably reduced—the pack-animals had forged their way ahead, and there were only five officers in the Pasha's suite; but we were preceded, accompanied, and followed by an escort of fifty cavaliers, whose Winchester repeaters were better calculated to inspire respect than Foreign Office passports or vizierial *bouyourouldus.*

Past the large village of Tirnowa and a cluster of other Mussulman villages nestling among maple - trees, mulberries, and vineyards at the foot of the Karadéré gorge, our route lay through the steep defile of Melouna to the first of the three mountain plateaux which rise, terrace-like, under the western flank of Mount Olympus, up to the central ridge of the Kambouni. Neither Kiepert's nor the Austrian staff map seems to convey a very correct impression of the orography of this important district. Far from being a confused alternation of hill and dale, it consists of three well-defined fan-shaped table-lands, rising one above the other like the steps of a gigantic staircase from the plain of Thessaly towards the angle formed by the *massif* of Mount Olympus and the long wall-like range of the Kambouni; and to each of these plateaux access can alone be obtained through one narrow break in the successive ridges of limestone rock which severally fence them in. Up the Melouna pass we had now reached the crest of the first of these ridges. To the south the whole plain of Larissa lay mapped out at our feet to the distant hills of Dhomoko and the blue outline of the mountains which form the present boundaries of the Hellenic kingdom—towns, villages, and farms dotted about in patches of dark verdure and white houses over the vast reach of ploughed fields and rich pasture-lands, broken only here and there by the glistening waters of a swampy lake or the silver thread of the Peneus meandering lazily across the flat. Turning northwards, we seemed to have a reflection of the same view reduced to a smaller scale,—a fertile table-land rich with a plenteous harvest of maize, which, owing to the difference of altitude, was here still waiting to be reaped; broad bands of red earth where the wheat crops had already been cut; villages half hidden amid groves of walnut-trees and willows; and at the farther extremity, under a rocky cliff such as that we had just scaled, the small town of Elassona. It is in this basin that the pure Wallach element begins. In Elassona itself there are, out of four or five hundred houses, perhaps one-third of Mussulmans; but in the other hamlets, and especially in the large neighbouring village of Tzaritzena, with its more than two hundred houses, the population is almost unmixed Wallach. Descendants they claim to be of the old Roman legionaries who colonised this region; and notwithstanding the influence of Hellenism, which has been during centuries of Turkish oppression their protector, their instructor, and their hope, both their language and their type are there to prove that they have no family affinity with the Hellenic Greeks. Hence, northwards into Macedonia pure Greeks are no longer to be found. All the communities which are included under that designation are Wallachs; or Romounoi, as they call themselves —Greco-Wallachs, as they are called by the Hellenes.

Elassona was to be our resting-place for the night, and we made straight for the house of an Albanian Bey whose hospitality Selami Pasha had bespoken. We had scarcely established ourselves in the *selamlik* before the interminable procession of officials, officers, and notables made their appearance in the room to pay homage to the distinguished guest. This ceremony was repeated wherever we halted, but I will not visit this daily infliction upon my readers. *Ab uno disce omnes.* First would appear the Governor or the President of the municipality, the civil "boss" of the place, in a very seedy official coat and Frankish boots of an extraordinary build; then the solemn Mufti, with a broad green turban, long white beard, and flowing garments; and the Kadi or judge, with a white turban and brown *kaftan*, generally obese and asthmatic, having grown fat on the pickings which belong to his profession; then the minor fry of officials, with their arms more devoutly folded and their backs more humbly bent as they descend in the official hierarchy; then the religious heads of the Christian communities, evidently rather ashamed at being caught by a foreigner in the act of cotooing to the infidel; and finally, the flood of notables, tall and short, stout and lean, dandified and dirty, Christian and Mussulman, all beaming with brotherly love and anxious to tread on each other's toes. Each one as he enters rushes up to the Pasha and pretends to kiss the hem of his coat or the toe of his boot, upon which the latter says, "Istaghfar Allah!" (God forbid!) and the disappointed visitor retires with becoming alacrity to squat down in the best place which may still be left on the divan, or, failing that, upon the floor. The Pasha then inquires how things are getting on in the village or town, and a chorus replies that prosperity cannot fail to attend the sunshine of his Excellency's presence. The answer seems generally to be regarded as entirely satisfactory, for the Pasha relapses into conversation with his companions, or singles out one of his visitors to talk more practical business, and the rest sit in meditative silence awaiting the advent of coffee.

This part of the performance depends for its success on the number of coffee-cups which the establishment can produce: if there are thirty visitors, and only three or four coffee-cups in the house, the relays are apt to drag. But no visitor dreams of leaving until he has had his cup of coffee, even as none think of remaining, except on special invitation, after having consumed their *quantum*. On this occasion, however, the solemnity of the proceedings was broken by the sudden irruption of a wild-looking individual with long hair and shaggy beard, whose naked limbs were wrapped in, but scarcely concealed by, a single dingy garment of coarse grey wool. He at once rushed up to the Pasha, kissed him on both cheeks, and, without asking by your leave or with your leave, proceeded to squat himself down between me and mine host. He was a Dervish, and Selami at once introduced him to me as belonging to the mendicant order of the Bedawi. He seemed inclined at first to scowl at the unbeliever; but when I told him that I had seen the tomb of the founder of his order, Sheik Ahmed el Bedawi, at Tanta in Egypt, and had visited his spiritual chief at

Damascus, his heart rapidly warmed towards me, and he apologised with touching simplicity for his rudeness. He had been brought up from his childhood in a *tekkeh* or monastery in Bulgaria; but during the Russian war the Christians of a neighbouring village had burned his *tekkeh* and murdered his father and mother, and he himself had been obliged to take refuge in flight. I somewhat warmly disclaimed, both in my own name and in that of Christians in general, all responsibility for such outrages; upon which he gravely took both my hands in his, and fixing his deep-sunk, fervent eyes on mine, bade me, if my words were true, welcome in the name of God. My words were true; but the earnest voice and searching look of the Dervish of Elassona have often haunted me since, both awake and sleeping, with their pitiful reproach.

After our visitors had been despatched, the evening meal was produced, and we soon dispersed to seek our several quarters for the night. Mine had been selected in one of the Christian houses of the town. But I was not yet to enjoy my rest. No sooner had I installed myself than the premises were invaded by the

Christian townsfolk, who evidently looked upon the jaded traveller as a mere machine for dispensing news, and were anxious to sound him upon certain points which they had not ventured to raise in the presence of the Turkish Pasha. The rumour had got abroad that I was a British Government official charged with the task of preparing for the transfer to Greece of the territories awarded to her by the Conference of Berlin. Though I vehemently disclaimed the invidious honour, my denial was received with scarcely disguised incredulity, and almost fierce were the queries as to how soon they were to become Greeks. But albeit the Hellenic influence was strengthened by the presence of two Greek priests, there was not, I confess, the general enthusiasm at the prospective change which I had been led in certain quarters to anticipate. The majority of my uninvited guests were Wallachs, and annexation to Greece meant in their eyes freedom from Turkish rule, and, so far, a blissful consummation. But the national feeling was strong among them. "We are a small nation," one of their spokesmen said, "and can scarcely hope for independence: therefore, better be dependent upon the Greeks, whose tongue we understand and whose faith we share, than on Turks, who are strangers and spoilers in the land; but Wallachs we shall always remain."

And this I hereafter found to be the prevailing feeling throughout the Wallach country. In the towns and larger villages, especially where priestly influence is strong, Hellenism is the fashion bred of preaching and of education; but it is not deeply rooted. Among the people, especially among the peasantry and the shepherd tribes, there is a strong national pride.

Among the mountain shepherds there is even a secret fear of annexation to Greece, which the Turkish authorities have not been slow to work upon. Most of them are nomads, who, as soon as the winter snows settle down upon the mountains, leave their summer alps and drive their flocks down into the plains of Thessaly. Were Thessaly in other hands than Turkish, they are afraid of being cut off from these pasture-lands, which would soon be redeemed for cultivation by more civilised and industrious owners. One of their shepherd patriarchs said to me—"God gives us the mountain alps for our flocks in summer, and in winter the Government gives us the waste lands of the plain. What more could the King of Greece do for us?" In the villages which have suffered more from Turkish oppression, it is the fear of military service, to which they are not liable under Turkish rule, which damps their Hellenic sympathies. And beneath all these motives of self-interest there lurks the consciousness of a distinct origin and vague national aspirations. At the present moment they are too poor and cowed to assert themselves. But should they receive encouragement and assistance from without, should the Roumanians be allowed time to push the propaganda which they have already initiated among them, I should not be surprised to see a Wallachian national movement spring up in these regions as strong and as unexpected, though perhaps just as little spontaneous, as, for instance, the Albanian movement. True, the whole power of the Greek clergy would be arrayed against it; but the Bulgarians have already shown that among the populations of this peninsula, notwithstanding ignorance and superstition, the pride of race, when once roused, bursts even the bonds of religious tutelage.

Dawn was just brightening into daylight as our horses clattered through the streets of Elassona and scaled the precipitous cliffs which rise like a wall over the last houses of the town. A narrow gorge, cleft by a mountain torrent, leads to the brow of this second ridge, and a more lovely view than that which awaited us on emerging from the ravine can scarcely be imagined. At our feet lay Elassona, its domes, minarets, and masses of green foliage still wrapped in the morning mist; over against us, on the other side of the defile, a quaint medieval monastery, said to be built on the foundations of an ancient Hellenic fortress; before us, another stretch of table-land studded with villages, and orchards, and fields of maize, with a double background of purple mountains; and close upon us, on the right, the giant buttresses of Olympus, rising almost sheer from the plain, with dark patches of forest hanging here and there on the flanks of the mountain, and its lofty domes, already crowned with the first snows of winter, glowing under the crimson rays of the rising sun. Descending into the little plain of Eleutherochoros, we began to meet Wallach shepherds already driving their flocks down towards the south,—wild-looking fellows, tall and gaunt, with long, often flaxen hair waving about their shoulders; heavy, bony faces; aquiline noses strongly developed and well cut; rather prominent hairless chins; dressed in strange fragmentary woollen kilts and leggings, with perhaps a cloak more tattered still, thrown back toga-like over the left shoulder, — each one surrounded by a little army of pigs, that run grunting and stumbling between their master's legs, while he lavishes upon his cherished brood alternate cursings and blessings in the choicest Wallach. More than the Bulgar, more even than the Montenegrin, does the Wallach love his pigs; never a Wallach village, never the merest hovel, without them; and when the shepherd wanders down to the pastures of the plain, with him they wander too, making music by the way.

From the plateau of Eleutherochoros another small pass across a ridge of hills leads into that of Wlacho Livada—a repetition on a somewhat larger scale of the two plateaux we had already crossed; the small town of Wlacho Livada, with nearly four hundred houses—the centre, as its name indicates, of this Wallach region—lying away to the north-east under Mount Chapka. In front of us the long chain of the Kambouni, varying between four and five thousand feet, marks the new line of frontier,—an unbroken wall of ironbound rocks, save for the deep depression which shows the entrance to the pass of Kirk Guetshid. The change of scene is sudden and abrupt. After riding for three hours across the park-like plain, the traveller enters a deep and gloomy ravine. Lofty cliffs of lime and sand stone rise on either side of the Sarantoporos to a height of 200 and 300 feet; mountain-ash rich with red berries, stunted oak, and a score of different shrubs, grow, as it were, out of the live rock; and alongside the torrent—which, already swollen with the first autumnal storms, leaps and foams over huge boulders of rock—a rough track, broken here and there by landslips, now winds beside the tortuous bed of the stream, now climbs along the face of the precipice. After about two miles the gorge gradually expands, and, rising over a succession of undulating meadow-lands, we reached the watershed, the frontier between Greece and Turkey that was proposed by the Berlin Conference. Forsooth, a strategic frontier!—behind us the gorge of Kirk Guetshid, and in front of us a sheer precipice falling away into the valley of the Vistritza and the plains of Macedonia. To descend from the mouth of the gorge is impossible; so we kept on along this lofty plateau for another two or three miles, and presently descried a cleft in the mountain-wall. The Turks call it Demir Kapou—the Iron Gate between Thessaly and Macedonia.

# CHAPTER IV.

## THROUGH WESTERN MACEDONIA.

THROUGH the massive limestone portals of the Iron Gate a steep zigzag, over which the ruins of a Venetian castle still mount guard, leads down to the Vistritza. The broad stream, which seems to carve for itself every year a fresh bed in the chalky soil, was easily forded, and on the opposite side a small party of horsemen were waiting to escort us to our night's quarters at a large Albanian *chiftlik*, belonging to a cousin of Selami Pasha. A short canter over undulating downs, intersected by deep seams of white chalk, brought us in half an hour to Pyrgos, a small hamlet consisting of some fifteen or twenty low mud-houses, amidst which stood out the solid square-built Greek chapel and the double-storeyed *pied à terre* of the Albanian Bey. The landlord, who happened to be spending a few weeks on his estate to superintend the payment of the tithes and the division of the year's harvests, was standing at the gate of the large courtyard ready to receive us and conduct us to our quarters. These were not, however, so easy to reach: they were situated in the upper part of the house, the basement being devoted entirely to storing provisions, grain, &c.; and the only access to our "residential flat" was up a rickety wooden ladder, which had already lost several bars under the ravages of time, while those that had been spared seemed to be the favourite roosting-place of a numerous family of hens. At last, notwithstanding the resentful crowing of a pugnacious cock, that step for step opposed the invading forces, we reached the spacious loft which had been fitted up for our reception. Bright carpets of many colours and piles of silk cushions contrasted strangely with the rough beams and rafters of the roof, black with smoke. But stranger still was the view from the windowless casements. A faint glimmer from the west still lighted up the mountain-wall over which we had descended from Thessaly, but the bed of the Vistritza and the intervening downs were veiled in a thick white mist. The courtyard beneath us was bright with half-a-dozen flaming fires, around which soldiers, muleteers, and peasants squatted in picturesque groups, busy with gigantic spits, and pots and pans, for the preparation of the evening meal; while others were hurrying to and fro, tethering up the horses, piling up the baggage, or doing swift execution on batches of unfortunate sheep and screaming poultry. Above the hum of voices, Turkish, Greek, Albanian, Wallach, sounded the hungry barking of the village dogs, disturbed by the unwonted invasion, and excited by the even more unwonted smell of roasting viands. In about an hour's time our host, who had been busy awhile among the crowd below, stimulating them by word and example, announced that supper was at last ready, and lifting the carpet hangings which alone separated the two apartments into which the loft had been divided, led us to the banquet-hall. The table—*i.e.*, a huge brass platter resting on a low stool about fifteen inches high—stood in the centre of the room, and around it we squatted on the floor in true oriental fashion. Six statuesque Albanian youths, retainers of the house, held flaming torches above our heads, while others climbed nimbly up and down the ladder-staircase, balancing on their heads the ponderous dishes which had been prepared below. It was a truly Gargantuan repast, both for the number and the dimensions of the courses. To a huge bowl of a kind of mulligatawny soup succeeded the obligatory whole roast sheep; and then, through a never-ending medley of *entrées*, sweets, roasts, more sweets and more *entrées*, we at last came to anchor on a pyramidal *pilaw*, the proper function of which in a Turkish dinner appears to be the filling up with rice of every nook and cranny that may yet have escaped the intrusion of more solid food, and the effectual checkmating of all digestive operations. Coffee, cigarettes, and ablutions were wholly unequal to dispel the lethargy superinduced by hard riding and harder eating, and we soon curled ourselves up in cloaks and quilts, to sleep, and snore, and dream away the *indigesta moles* of our feast.

The sun was already streaming over the silver

domes of Mount Olympus when our host threw open the wooden shutters of our casement and let in the crisp morning breeze; but as we had only a short day's work before us, there was no need to hurry. So, while the horses were being leisurely saddled, I sauntered over the *chiftlik* of Pyrgos. It gave its name to one of five similar estates belonging jointly to three Albanian brothers. The lands attached to it consisted of about 2000 *donoums*, or 1400 acres. These lands were farmed, as in Thessaly, on the *métayer* system, and provided work and occupation for fifteen families, whose houses composed the *chiftlik*. There were twenty-five adult male labourers, and altogether one hundred and twelve souls on the estate. Thirty-five pairs of oxen, the property of the *chiftgis* or peasantry, were used for farming purposes, the whole of the land being devoted to agriculture. The soil, though light, was generally fertile, as there was nowhere any lack of water: barley, wheat, oats, cotton-wool, and Indian corn were grown in the usual rotation, one acre in three being allowed to lie fallow every year, in order to prevent the exhaustion of the soil. Manure was never used, except as fuel. Yet the landlords' share of the profits, after deductions for tithes and taxes, generally exceeded 1000 *liras* a-year; and as the same amount accrues under the *métayer* system to the farmers, the share of each household (consisting of seven souls) could not be set down at less than sixty *liras* (£54), —an income which, considering their modest wants and the cheapness of all necessaries of life, certainly represents a larger measure of comfort and prosperity than falls to the lot of many other groups of peasantry living under the paternal rule of more civilised Governments. Nor are these exceptional or specially selected figures. Throughout this part of Macedonia I found the farmers on the various estates which I visited very evenly circumstanced, and satisfied, as they well might be, with their condition. Much, of course, depends upon the personal character of the landlord or agent; but the interest of the latter alone operates effectually to check any tendency to short-sighted oppression. Christian villages are as a rule more flourishing than Moslem villages, as they escape by the payment of a comparatively trifling poll-tax the curse of the conscription, which has of late years entirely drained some of the latter of all their adult male labourers.

From Pyrgos we made a short detour to visit a *chiftlik* belonging to Selami Pasha himself,—a pretty little hamlet ensconced among walnut-trees and apple-orchards on the banks of a bright, bubbling stream. Our visit was entirely unexpected, and almost all the population had turned out to work in the fields; but the Pasha was soon surrounded by a small host of children, who kissed his hands and curtseyed in the most approved fashion at the bidding of an ancient village priest. But the news of our arrival was soon bruited about, and presently the peasants came flocking in to do homage to their lord with much wasting of gunpowder. Every one, too, had his own tale to tell and his own grievance to set forth, and to every one the Pasha listened with unvarying good-humour, giving the agent, who was in attendance, orders to remedy such complaints as appeared reasonable, and dismissing the more futile petitioners with a gentle joke, to which the bystanders never failed to rise. It was evident that here at least the peasants did not look upon themselves as the victims of grinding oppression. Village justice having been dispensed under the walnut-tree, we were free to devote the rest of the morning to pleasure, and the Pasha ordained that pleasure should take the form of sport; for, on the strength of two handsome pointers—who, however, never pointed—and of a gorgeous double-barrelled breech-loader rich with gold and silver chasings, his Excellency considered himself to be a mighty hunter before the Lord. Unfortunately, the hares and partridges were always too quick or too slow for him, and obstinately declined to get in the way of his shot: so, although there was no lack of game, the only trophies which we finally brought back from our expedition were two pigeons which the Pasha carefully stalked and potted in an olive-tree, and a healthy appetite for our noonday meal.

In the afternoon a few hours' ride over undulating uplands studded with Wallach villages, and vineyards still rich with grapes, brought us to Kosana, a pleasant little city of some 7000 souls, celebrated for its excellent vintages. The wines of Kosana are perhaps better known to English readers than they wot of; for it is one of the *grands crûs* of Macedonia, and during the last few years a considerable quantity has been exported *vid* Salonica to Bordeaux, where it is converted, by the simple process of diluting, bottling, and labelling, into "light and wholesome" claret for the British market. The pack-animals and the greater part of the escort had been despatched in the morning direct to Kosana; and the good people of the town, thus forewarned of the Pasha's approach, had had full time to make preparations for his reception. About a mile from the town the officials and municipality and a detachment of the garrison were drawn up in solemn array, and a little farther on the Greek Archbishop and three ecclesiastical dignitaries of his household came ambling gently towards us on confidential mules. His reverence was an old friend of the Pasha; and as soon as the latter caught sight of him, he dismounted to receive him with proper respect. To have witnessed the cordial meeting of the two kindly old gentlemen, as they stood hugging each other in the muddy road, and lavishing on each other all the pious formulas of the Christian and Mussulman greeting, would have strangely astonished our holy fanatics at home, whose charity endureth all things, save Turks. We were to enjoy the bishop's hospitality for the night; and our motley cavalcade soon threaded its way through the narrow streets of the town to the quaint old episcopal palace, where Greek maidens besprinkled us with rose-water from the overhanging windows as we rode through the gate into the spacious courtyard. The palace ran round three sides of this court, which was overlooked by a double arcade, the lower one resting on handsome Corinthian columns, the spoils of some ancient edifice; while the fourth side opened into a bright little garden, beyond which the view extended over vineyards and fields to the blue line of distant mountains over against Thessaly.

Kosana is the last purely Wlacho-Greek town in Macedonia. Northwards the Wallachs are henceforth only found interspersed with Albanians, Turks, and Bulgarians, and generally in a minority. Here they are still unmixed with any other nationality; and on the wall of the bishop's reception-room there hangs an ancient map, dating from the last century, on which Kosana is marked in large letters as one of the chief towns of Roumania, that name being applied not to Moldo Wallachia, but to the region I was just crossing from Larissa to the Lake of Ostrovo, twenty miles south of Monastir. But now the name is contemned or forgotten; and though in the 900 houses of this city there are scarcely twenty where around the family fireside any other language is spoken than the old Latin-sounding Wallach, the prosperous townsfolk would be deeply hurt if any doubt were hinted as to the genuineness of their Hellenism. For clerical influence is strong here—strong not only with the strength of ecclesiastical authority, but with that greater strength which it derives from the devotion of the population to the venerable old prelate who has lived and done good amongst them for upwards of forty years. An exception among his class, he has Hellenised his flock not only by schools and sermons, but by kindness and uprightness. The overbearing grasping character of the Greek clergy has too often undone the work of Hellenisation wrought in the schoolroom and the pulpit. Here the contrary has been the case; and there is therefore little cause for wonder if in the whole Wallach region there is no stancher bulwark of Hellenism than Kosana. But the Hellenism of its inhabitants is not of the heroic kind. Like most towns in Turkey which are entirely or preponderatingly Christian, they have had little to suffer from Turkish rule. A Turkish official generally finds that it pays best to be on good terms with the majority of the population; and so, where the Christians largely preponderate, he makes friends for the nonce with the mammon of unbelief. In Kosana it is the bishopric which really governs the town; and the authorities are well content that it should be so. For the inhabitants pay their taxes regularly, with sometimes a little *douceur* over and above; and if their sympathies are Panhellenic, they always restrain them within the most Platonic bounds. So little, in fact, does the Government fear intrigues or disturbances on their part, that it has lately distributed arms among the population of the town and the surrounding villages to defend themselves against the brigands. A more united family, the bishop assured me, could not be found, than the medley of priests and officials, Christian notables and Moslem divines, who gathered in the evening round the episcopal table, and drowned their sectarian differences in deep draughts of Kosana wine. Under its influence even my neighbour—an ascetic and toothless Archimandrite, who at first evidently viewed the proceedings with grave doubts as to their orthodoxy, and occasionally relieved his mind by whispering with a grim sardonic smile the name of Gladstone into my ear, as if it were a charm to exorcise evil spirits—relaxed into a more jovial humour, and, after the customary toasts had been disposed of, insisted upon drinking the health of every guest present in separate bumpers.

Rain was falling in torrents the next morning, and it was decided to postpone our start till noon. But even before that hour the weather had cleared up, and the sun gleamed through the clouds, lighting up the bright groups of quiet townsfolk who thronged the streets as we rode out of the town; for it was Sunday, and church-time being over, there was nothing for the Kosanlis to do but to loiter about the streets and show off their holiday attire,—a task of love, which the gentler sex especially performed with much success, one dark-eyed damsel fairly taking my heart by storm, as, whipping off her little red high-peaked shoes in order not to soil them in the puddles, she tripped across the street and pelted me with rose-leaves. Outside the Greek schoolhouse the children of both sexes were drawn up in full battle array, under the command of masters and mistresses, and greeted us as we passed with the Sultan's anthem,—a healthy, bright-faced little army, whose joyous voices could still be heard carolling behind us as we wound over the hills, through acres and acres of vineyards. The country which we now traversed was an undulating table-land, flanked on either side by ranges of hills running north and south perpendicularly to the frontier chain of the Kambouni. The fields looked fairly well cultivated, each being divided off from its neighbour by neatly-piled-up walls of loose stones. Small hamlets, which a minaret or the whitewashed dome of a *veli* generally marked as Turkish, nestled among trees in the shallow depressions of the plateau; and here and there a ruinous tower recalled the days when the *derebeys* held lawless sway over the land. Four hours' heavy riding, over tortuous tracks which the night's rain had converted into quagmires, brought us to our destination,—a large Mussulman village called Kaïlar—or rather, as its plural termination indicates, two villages grown into one. Kaïlar has made no special mark in history; but it is not without a certain fame in the surrounding country, and I, or rather my horse, felt a peculiar interest to reach it—for the unfortunate animal had been badly shod before leaving Larissa; and whenever I remonstrated on the subject with Hassan Tchaoush, whom Selami had told off to be my groom-in-waiting during the journey, that worthy sergeant had invariably answered: "Ah, Effendi! just wait till we get to Kaïlar: that's the place to have the horses shod. *Wallah!* there are no smiths in the world like unto the smiths of Kaïlar." Many towns and districts in Turkey thus have their speciality. The Albanians of Dibra are reputed as wood-cutters; the Bulgarians of Kezanli are celebrated gardeners; monastic Zitza furnishes the country with *hanjis* or ostlers; Kaïlar, it seems, is renowned for its smiths. Whether they be worthy of their fame is a point on which I must refer the reader to my faithful Rosinante.

The next morning we were off betimes; and in order to make up for the half-day we had lost at Kosana owing to the rain, we had to give up making the detour to Wlacho-Klissura, which Selami Pasha had originally proposed to take on our way, in order to enable me to visit the centre of the Roumanian propaganda among the Wallachs of Macedonia. Our route lay, as on the preceding day, over narrow undulating plains, flanked on either side by mountains; and, save that we were entering the Bulgarian region, and that, as we progressed northward, Bulgarian villages steadily preponderated over Turkish and Wallach, there were few features of interest to mark the journey. About three hours from Kaïlar, the Lake of Ostrovo came into sight. A century ago, fields and meadows and flourishing villages were to be seen where now lies a long sheet of dark-blue waters, reflecting, like a mirror, the barren slopes of Mount Bermius, which overhang its eastern shores. The long plateau which we had followed ever since leaving Kosana, sinks steadily towards the north until it is barred by a ridge of hills running across from east to west. The depression thus formed was formerly drained by a stream, which lost itself, no one knew whither, underground. But an earthquake, or some other cause, suddenly blocked up the channel, and the waters which ran down from the surrounding heights, failing to find an exit, accumulated in the hollow, and covering acre after acre of ground, and swallowing up hamlets and villages, have finally formed a lake, which is now nearly twelve miles in length and two in breadth. It is still slowly rising, inch by inch and year by year; and unless it can once more force open a subterranean passage for its waters, it must inevitably continue to rise until it reaches the level of the plateau which divides it off from the basin of the Vistritza. How many more prosperous homesteads, how many more acres of fertile land, will ere then be buried under the placid surface of its relentless waters!

After halting for our mid-day rest at a pretty

*kiosk* built by the late Abdi Pasha on the banks of the small lake of Petersgrad, we paid a visit to the large Bulgarian *chiftlik*, where the eccentric old Turk used to come down every year and superintend, with much joyous carousing, the making of his wines. Ten gigantic casks, each holding 1000 *okes* of wine, were annually filled with the best vintages for the use of his own house. Over the crest of a low range of hills, we at last began to descend into the plain of Monastir; but though the lofty peak of Mount Peristeri marked the spot where the city lay, the day was too far advanced for us to reach it before nightfall. So, winding along the spurs of the eastern chain of mountains which flank the plain, we brought up for the night at the pretty little village of Raduwicz, the property of the three Albanian brothers whose hospitality we had already enjoyed at Pyrgos after our second day's journey. Raduwicz occupies, according to local tradition, the site of an ancient Greek town; but besides the fact that there never can have been room for a town in the narrow valley where it lies, cramped up between two spurs of Mount Nidje, no place of importance appears to be mentioned in this locality. Possibly the tradition has been invented to enhance the value of one or two Greek relics which have been unearthed by the villagers. The headman of Raduwicz insisted that I should see the *anticas.* No European had ever yet visited Raduwicz, and I was therefore bound to give my opinion on them. So I trudged with my guide up the hillside to a small Bulgarian church, where I was shown a stone let into the pavement, which bore a Greek inscription. It was, however, so worn by the long traffic of pious feet, that I was quite unable to decipher it; but I saw that the stone had been considerably mutilated in order to fit it into its place. A fossil *papas,* who was kind enough to try and throw some light on the subject by holding the altar lamp over me, assured me that it was an excellent stone for exorcising evil spirits, but that it was useless to attempt to read the inscription, as it was written in the devil's language. So I retired. The other *antica* was a medallion let into the wall of a farmyard, showing the heads of two youths in relief, with the fragmentary inscription *του κυκυλλου*; but the marksmen of Raduwicz had evidently selected the unoffending youths as a target, and their faces were so disfigured that it was impossible to judge of the artistic value of the workmanship.

Only a faint streak of light over the eastern hills heralded the approach of day as we set out the next morning to complete the last *étape* of our journey; but the sun was already high in the heavens before we reached the imperial road, which leads, or ought to lead, were it only completed in the middle, from Salonica to Monastir, and following the broad avenue of poplars, soon found ourselves in the large square of stately barracks which mount guard over the entrance to the city.

# CHAPTER V.

## THE VILAYET OF MONASTIR.

A PRETTY little city is Monastir. At the foot of the defile which leads from Macedonia into central Albania, backed by bold and lofty mountains already capped with winter snows, it lies half buried amid foliage in an emerald-green valley; graceful domes and minarets, stately barracks, and clusters of bright white houses, hedged in with dark-green walnut-trees and glistening silver poplars and orchards rich with all the varied hues of autumn, while bubbling streams fresh from their alpine homes leap merrily through its sunny streets and picturesque bazaars. In its crowded thoroughfares may be met all the motley costumes of European Turkey,—the fair-haired Bulgarian enveloped in his shapeless grey cloak and hood, the gaunt slouching Wallach shepherd in his grimy *fustanella*, the proud Albanian highlander bristling with pistols and daggers, the squalid Iberian Jew, the sharp-eyed Greek merchant in seedy European clothes, the solemn beturbaned Turk in flowing *kaftan;* for Monastir—or, as the Christian prefers to call it, Bitolia—is the meeting-point of many nations. As on most Macedonian cities, the curse of many tongues, and many races, and many creeds, has fallen heavily upon it. With a population which certainly does not exceed 35,000 souls, and which some put down at little over 25,000, it boasts nearly a dozen different communities, with separate aspirations, separate interests, and separate hatreds. There are Bulgarians who speak Bulgarian, and have joined the Bulgarian schism; there are Bulgarians who speak Bulgarian and share the political aspirations of the schismatics, but have not yet thrown off their allegiance to the Greek Patriarchate; finally, there are Bulgarians who have been so far Hellenised as to speak the Greek tongue in preference to their own, and profess the Hellenic creed, both religious and political—but their number is small, and daily diminishing. Then there are the Greeks, who are, it is true, Greek only in name; for, with the exception of the archbishop and the Hellenic consul, there is scarcely a family in Monastir that can lay claim to pure Greek blood. They are Wallachs; and among them, again, there is a large and growing section which still speaks its own tongue in preference to Greek, and has distinct national aspirations outside of Hellenism. The Mussulman community is scarcely more homogeneous: there are Albanians who still hold with the Porte, but the greater number look only to its support as the means to an end—the vindication of their own independence. Monastir especially is eloquent with memories of Turkish treachery scarcely calculated to stimulate the devotion of the Albanians to the empire. It was here that in 1830, after the Greek war of independence, Reshid Pasha invited some 500 Albanian beys from northern and southern Albania to a banquet to celebrate the amnesty proclaimed by the Sultan; and when they rose to leave the vizierial tent, two battalions of Turkish troops were drawn up to give them military honours: but instead of the order to present arms, the Grand Vizier gave out of his own mouth the word of command to fire. In a few moments the meadow, which the Albanians still call "The Field of Treachery," was red with the best and noblest blood of Albania. Never an Arnaut escaped from the foul ambush where Albanian independence was laid prostrate until the ferment of recent events galvanised it once more into life. The Turks themselves may be subdivided into two categories; the migratory and predatory officials—and the settled working classes, artisans, merchants, husbandmen, &c., who, perhaps more than other sections of the population, have what M. Gambetta calls the passion of peace, and yet are probably destined henceforth to be the first victims of war.

To obtain anything like accurate information as to the exact numbers of these various communities is a task of extreme difficulty, the art of cooking statistics having been carried here of late wellnigh to perfection. As an instance thereof I subjoin the statistics of the population of the city of Monastir, obtained from three authorised sources :—

GREEK RETURNS.

| | | |
|---|---|---|
| Greek houses, | | 2156 |
| Bulgarian „ | | 529 |
| Mussulman „ | (about) | 2700 |
| Jews' „ | (about) | 800 |
| | | 6185 |

BULGARIAN RETURNS (approximate).

| | |
|---|---|
| Greek houses, | 1200 |
| Bulgarian „ | 2100 |
| Mussulman „ | 2400 |
| Jews' „ | 700 |
| | 6400 |

TURKISH RETURNS.

| | |
|---|---|
| Christian houses (Greek and Bulgarian), | 2224 |
| Turkish „ | 2327 |
| Jews' „ | 625 |
| | 5176 |

The Turkish returns unfortunately amalgamate Greeks and Bulgarians, the Government distinguishing only between Christians who pay the military-exemption tax, and Mussulmans who are liable to active service. These statistics may, however, be generally trusted as furnishing the minima of the two communities, each being interested in reducing the returns as far as possible in order to elude on the one hand conscription, and on the other the payment of the exemption-fee. The religious denomination forms the basis of the Greek returns, the language spoken that of the Bulgarian returns, each being selected from a party point of view, with the object of establishing a claim to political supremacy. But of the two, there is little doubt that the latter are the least deceptive; for there are few Bulgarophones who do not share Bulgarian national aspirations, while among those who still profess the orthodox Greek faith, most of the Bulgars and many of the Wallachs and Albanians have already abjured the political creed of their spiritual teachers. From various calculations, I have arrived at the conclusion that the population at Monastir, apart from the Jews, may be divided into three nearly equal sections, the Mussulmans being slightly over one-third, another third being composed of Hellenised Wallachs and Bulgars, and rather less than a third consisting of Bulgarians. But take the whole *vilayet* of Monastir, or go further and take the whole province of Macedonia—namely, the four *sandjaks* of Monastir, Salonica, Drama, and Serres—and the proportions obtained will be very different. Monastir is, at most, only a detached fort of Panhellenism. After leaving Kosana we lost sight completely of the Greco-Wallach element, all the villages along our road being either Turkish or Bulgarian; and the farther north or east one goes from here, the stronger the Bulgarian element becomes. To the west the confines of Macedonia are close at hand; and yet even in the mountains which, bounding it in that direction, are the bulwarks of Albania, Bulgarians are to be met with in large numbers even as far as the Lake of Ochrida.

The most striking, and at the same time the most melancholy, feature in Macedonia, is the bitter hostility of the Christian communities. Fanaticism no longer runs in the old groove—Christian *versus* Mussulman. The old enemy is moribund, and the young races are making ready to fight over his inheritance. The Congress of Berlin may have seen fit to overrule the Treaty of San Stefano, but in the eyes of the populations of Macedonia it is still a living writing on the wall—to the Turks a sentence of exile and of death, to the Bulgarians the charter of their rights, to the Greeks and other Hellenising races a perpetual menace. The two rival

Christian races know the change is at hand; but how will it be finally shaped? Is Hellenic civilisation, which, by its traditions, its schools, its ecclesiastical discipline, its influence with the conquerors themselves, has kept alive during so many centuries of servitude the sacred fire of liberty among the Christian populations of the peninsula, to be sacrificed again to the uncouth despotism of northern barbarians, whose only strength is in their numbers and the friendship of a powerful neighbour? Are the Greeks, answers the Bulgarian—are a small minority in the land, who have traded ever since the invasion on the supineness of the conqueror, in order not only to share with him our spoils, but to impose upon us the arrogant rule of their alien prelates and alien schoolmasters—to rob us at the eleventh hour of our birthright, and substitute for the haphazard misrule of the Turk the tyranny of a crafty priesthood and its servile nominees? It is not against the Turk that his worst invectives are now directed. Turkish officials are still too often unjust and grasping, as in the past; the courts are still corrupt; the hand of the Mussulman bey may still be heavy on the unbelieving rayah,—but the Bulgarian now detects in them qualities to which he was formerly purblind. He has a good word to say for the natural kindliness and hospitality of the rich, for the courage and patience of the soldier, for the industry and good-heartedness of the peasant. Officials and landowners are not all as black as he used to paint them; he finds excuses for their faults—ay, even for their crimes. His most bitter denunciations are reserved for the Greek. For he is the rival whom he now most fears, and therefore most does hate. *Mutatis mutandis*, the Greek expresses in the same terms his hatred of the Bulgar, both agreeing that even Turkish rule is less intolerable than would be that of the accursed rival.

The Greek clergy has a heavy responsibility to bear for this disastrous estrangement, which is principally the outcome of its own intrigues. Using and abusing the power which he has hitherto wielded under Turkish rule as the recognised head of the Christian communities of the empire, the Greek bishop has too often worked not to cement the union of the various races of his flock in the fellowship of a common faith, but to promote, if not the meaner ends of personal ambition, at all events the narrow interest of a politico-ecclesiastical propaganda. The sudden awakening of the Bulgarian nation, the formation of a Bulgarian Church, were rude lessons, pregnant with instructive warnings. But he has failed to conceive their meaning. Angry at the past, fearful for the future, he tries in vain to stem the torrent which has already carried away so much of his laborious edifice, and, like Xerxes of old, would fain flog it into obedience. During the Russian war, when the Turks were only too prone to angry suspicion, the Greeks were the first to inflame their most dangerous passions by denouncing to them as Russian *spies every Bulgarian who had joined or* was suspected of joining the national Church. It was at the instance of the Greek Archbishop of Monastir that the Turkish Governor-General seized all the Bulgarian school-books in the *vilayet;* it was at his solicitation that the schismatic priests were driven out of the country, and that to the present day the Bulgarian Exarch of Monastir remains in semi-exile at Sophia, and the Bulgarian Church in this province is still disorganised, without a proper head or a proper constitution. Since the Berlin Congress gave Turkish rule in these regions a fresh lease of existence, the Greek clergy has redoubled its efforts to regain its wavering hold over the Christian populations. The powers placed in its hands by the constitution of the empire are strained at every point by unscrupulous prelates, whom the authorities are either too careless or too tolerant to check. Greek priests are forced upon Bulgarian villages, Greek schools are multiplied all over the country, and the *zaptieh's* aid is enlisted in order to compel the attendance of Bulgarian children, who are required to swell Panhellenic statistics. Even brigandage is impressed into the service of the Orthodox Church. But all is of no avail. In spite of menaces, and deeds by which these menaces are too often translated into reality, every day fresh bodies of Bulgarians, fresh villages, drive out their Orthodox priests and join the national schism. Many of them do so, indeed, with a sorry heart —excommunication has not yet lost all its terrors in this country. "But if we must choose," they say, "between the Patriarchate and our nation, our choice is already made. Let them make us schismatic; we appeal to the mercy of our Lord God." And as soon as the new organic statute or any other *régime* is introduced which shall allow the Bulgarian Church to be fully organised according to the rights conferred upon it by the imperial firman, the whole Bulgarian population will secede *en masse*, and the already waning power of the Greek clergy, sapped by its own short-sighted policy, will fall ignominiously to the ground.

The virulence of sectarian feeling, pregnant with the gravest dangers for the future of these regions, has already borne its baneful fruit in the growth and development of an evil, for which it would be difficult to find a parallel without going back to the days when the *Freischaaren* desolated Germany at the close of the Thirty Years' War.

On both sides of the Vardar, from the borders of the principality of Bulgaria to the highlands of Albania and the mountain-chains of the Kambouni and Olympus, bands of brigands, thirty, fifty, and a hundred strong, drilled and armed with the best modern weapons, sweep over the country, rifling and burning villages, carrying off prisoners for ransom, committing the most savage excesses, shedding blood in wanton cruelty, rendering the highways and byways of the province impracticable, and driving terror even into the heart of the towns. And the evil is steadily on the increase. Tempted by impunity, frightened by threats, driven by despair, peasants are every day leaving their plough to take to the mountains. The fact is, brigandage here is the outgrowth not only of social disorders, but of the political situation. A series of bad crops, and consequent dearth, which last year amounted almost to famine; the misery bred of the last war; the stimulus given to all the worst passions of an ignorant population by the exacerbation of political and religious rancour; the inefficiency of the police, whom want of pay and of discipline drive into complicity with the law-breakers; the insufficiency of the army available for garrison duty; the apathy or powerlessness of the local authorities, whose best intentions are frustrated by the constant blood-sucking of Constantinople,—have co-operated to engender brigandage: but it is by colouring his pursuit with patriotic pretences, by tempering his excesses with political considerations, that the brigand has succeeded in securing not only impunity, but popularity and admiration. In the south-western districts of Macedonia the bands are almost exclusively Greek; and by sparing to a certain extent the peasantry and villages who adhere to the Orthodox creed, religious and political, they claim, as the successors of the Clephts and Armatoles of the war of independence, the sympathies of the Greek populations. *Facile princeps* among these bands is that of Kathrakia, which carries its depredations from the foot of Olympus to the sources of the Vardar. Along the road from Wlacho Livada to Monastir there was scarcely a Mussulman, Wallach, or Bulgarian village which had not some tale to tell of either threats or deeds of violence. *Facile princeps* also is Kathrakia in cold-blooded cruelty, in proof whereof the two following incidents may suffice: At Khroupista, a large mixed village about a day's journey from Kosana, Kathrakia arrived with his band on the night of the 30th of October, and attacked the house of the Mufti, which is in the outskirts of the village. The Mufti was away; but in the house there was a considerable sum of money, nearly 3000 *liras*, the proceeds of a sale of land. Of this handsome booty Kathrakia promptly possessed himself; and then, without even the excuse of baffled greed, proceeded to the most hideous excesses on the three unfortunate children of the Mufti, who, with an old servant, were the only occupants of the house. The bodies of the two boys and a girl, seven, nine, and twelve years old respectively, were found the next morning cut into pieces, and showing traces of the most shameful maltreatment.

About seven hours from Monastir there is a small and prosperous Wlacho Bulgarian village, Melovishta. One Sunday morning the inhabitants were collected in church when Kathrakia and his band swooped down upon the place, leaving most of his men posted outside the church. The captain—who is, I believe, a Greek deserter—all gorgeous in the dress of a pseudo-Albanian chieftain, covered with gold, marched into the sacred edifice, followed by an equally gorgeous aide-de-camp carrying his Martini rifle, and a dozen men armed to the teeth. While divine service was proceeding he listened with devout reverence, and received on bended knee the blessing with which at the end of Mass the priest discharges his congregation. Then, wheeling round upon them, Kathrakia informed the terror-stricken villagers that they were his prisoners, and that he was about to select some eight of their number who were to follow him as hostages into the mountains until such time as their friends should send and ransom them with 3000 *liras* in gold. The peasants were then marched out between the brigand ranks, and Kathrakia having made his choice of eight —men, women, and children—his horsemen took them up behind them, and the band scampered away. In the course of about ten days the unfortunate villagers of Melovishta managed to collect some 200 *liras*, which two young men, who had near relatives among the hostages, volunteered to carry to the rendezvous appointed by the brigand chief. After meeting some of Kathrakia's scouts, and being marched about for a couple of days, they reached the headquarters of the band, and counted out their *liras* into the hands of the dashing *kapitanos*. But the bribe was too mean for him, and casting the money back to them, " Go back," he said, " to your people, and take your paltry money. I am not to be thus trifled with. Not one *piastre* will I abate from the sum I have demanded. I have kept my part of the pact until now. My prisoners are well and safe. But beware how you shirk your share of the conditions." Thereupon he ordered the prisoners to be paraded before the village envoys; and as the last one of the eight, an old man, was passing, Kathrakia out with his sword, and the unfortunate peasant's head rolled at the very feet of his horror-stricken son and nephew.

" Go now," Kathrakia said, "and tell the good folk of Melovishta what you have seen. They had better not again doubt my earnestness, or dare to keep me waiting." The two villagers arrived while I was still at Monastir, with a deputation from Melovishta, to plead for help from the Government.

In other cases Kathrakia appears yet more distinctly in the character of a defender of the true Orthodox faith, plundering Bulgarian villages and murdering schismatic priests. I was shown, while at Monastir, a letter addressed by him to the Bulgarising villagers of Derab. It was signed by him and his aide-de-camp, and bore the imposing seal of the Kapitanos Anasthasios Kathrakia. It began, "Dogs, hounds, sons of hell," &c., and proceeded to the effect that news had reached the ears of the mighty *kapitanos* that the villages of Derab harboured in their midst a Bulgarian priest who would fain lead them in the path of a pestilent schism. "Be it therefore known unto you," it concluded, "that unless you forthwith eject from among you the accursed brood of Satan, your village shall be laid low and yourselves put to the edge of the sword."

If Kathrakia's band is the most formidable of all and the most ubiquitous, there are many lesser stars who would fain rival him in their respective spheres. Niko has restricted himself to smaller exploits since the great *coup* he landed by Colonel Synge's capture. But Arcadi, Kriko, and a score of other names, spread terror and desolation in their several localities. North-west are the Albanian bands of Dubr, the black sheep of their nation, who seem to practise murder and pillage indiscriminately on all races and sects and classes, and hold the mountains all along the western edge of the Monastir plain up to Krichevo and the upper valley of the Black Drin—demons who have the fanaticism of crime, and defy even the authority of the League. To the east Bulgarian marauders render the roads unsafe, and commit here and there desperate excesses; but none of them west of the Vardar seem to have attained the high state of organisation and discipline which distinguishes the Greek bands.

When brigandage is thus rampant throughout the land, and has grown almost into a recognised institution which defies all rights and laws, it is not surprising that crime of every kind should be rife, and the prevailing anarchy turned to account for the satisfaction of individual passions and resentments. And while life is thus rendered intolerable for all classes of the population — Christian and Mussulman, Bulgarian and Greek, Albanian and Turk—while lawlessness has reached such a pitch as to rouse from their supineness the higher Turkish authorities of the province, Constantinople goes on draining Macedonia of the men and money without which the best will in the world is powerless to wrestle with these disorders. So long as justice can be bought in court by the worst culprits; so long as the police and minor officials, not to speak of the higher authorities, are driven through sheer want into collusion with the brigands; so long as, owing to general misrule, the sympathies of whole sections of the population are enlisted on behalf of the brigands; so long as there is no strong and just Government to check the indifference, not to say gratification, with which crimes committed on rival races or communities are nowadays viewed,—so long will the evil go on growing apace until the final catastrophe; and signs are not wanting that that catastrophe is already close at hand.

On the farther banks of the Vardar the Bulgarian revolution is already organised. It has in all the Bulgarian districts its committees who carry on its propaganda, and its bodies of avengers who carry out its behests. Formed to combat Hellenism by encouraging national education and promoting the interests of the national clergy, these committees are equally directed against the Mussulman landowners, and have taken against them the shape of a kind of Land League. Their object is, in theory, whenever a deed of oppression or injustice is laid to the charge of a Moslem, to visit upon him the punishment which no Turkish court of law can ever be found to inflict. But in reality their practice far exceeds even this code of savage justice, and seems more calculated, by fomenting disturbances and provoking reprisals, to give rise to an outbreak of fanaticism such as that which in 1876 precipitated the downfall of Turkish rule in Roumelia. Though they can be seldom accused of acts of plunder, these bodies of avengers wreak their cruelty and lust indifferently on the guilty and the innocent; and women, and children, and old men have to suffer in memory of ancient wrongs.

# CHAPTER VI.

## A BULGARIAN CITY.

Thanks to Selami's friendly convoy, I had reached Monastir pleasantly and safely. The question now presented itself of how I was to return to Larissa, and its solution was beset with the same difficulties which had met my desire to leave it. The most interesting route was without doubt to follow the direct road to Salonica through Vodhena to Yenidjeh, the royal city of the Macedonian kings and the birthplace of Alexander; and thence, skirting the eastern slopes of Mount Olympus along the shores of the Ægean, to re-enter Thessaly through the Vale of Tempe. But Ahmed Eyoub Pasha, who was then Military and Civil Governor of Western Macedonia, pleaded brigandage as an excuse to place an absolute veto on that route, the real reason for which ought probably to be sought in certain earthworks which were then being erected along some portions of that road. I was too much indebted to Selami's kindness to be free to act in direct defiance of the Turkish authorities; and so there was nothing left for me but to conform to the Vali's wishes, and follow the ordinary and prosaic route, joining the Mitrovitza-Salonica Railway at Graczko, whence (inshallah!) the *vapor* would carry me in a few hours to Salonica, and I could there (inshallah!) catch another *vapor* for Volo. That steam locomotion was not to my mind, and that I had other objects in travelling than to be whisked from place to place with a minimum of physical exertion, were ideas far too incongruous to the Vali's nature—which had already too often asserted on the battle-field its antipathy to all forms of locomotion save, perhaps, that which is popularly ascribed to the crab — for me to question the transcendent advantages of the *vapor*, especially when enhanced by the promise of a free pass, with which his Excellency was good enough to clench the argument.

The sun was just gilding the domes and minarets of Monastir, and glinting through the foliage of the trees, already thinned by autumn winds and frosts, as I rode out of the northern gates of the town. The lofty table-land, studded with tree-girt villages, lay stretched out before us, a broad reach of fertile arable land, broken only here and there by the silver streak of an undrained morass. A good road runs across this plain to Perlepe, which lies at the eastern extremity of the plateau, at the mouth of the two gorges which lead through the chain of the Babona mountains to the valley of the Vardar; and the bell of a quaint old clock-tower in the market-place of Perlepe was only just striking six *à la Turque* (about noon, European time) as we cantered across the bridge which spans one of the headwaters of the ancient Lydias.

As, coming up south from Thessaly, Kosana was the last town we passed where the Greco-Wallach element was unmixed, and beyond which we scarcely again heard either Wallach or Greek, except in the Greek quarter of Monastir, so, proceeding northwards from Monastir, Perlepe is the first large Bulgarian centre where the Bulgarian tongue is almost exclusively spoken. Even the Mussulmans of the town, who form about a quarter of the population, know Bulgarian as well as Turkish, and only speak the latter in official circles or among their families; while the Moslem villagers of the district often speak the Slavonic tongue in preference to Turkish. Nor has Bulgarian much to fear from the competition of Greek or Wallach, for there are now only eighty houses left here where either of those tongues is still spoken. Perlepe affords one of the most striking instances of the rapid strides with which the Bulgarian national feeling is outgrowing both Turkish domination and the Hellenising schemes of the Orthodox clergy. In the elaborate returns of the Greek population in Macedonia and Thrace which the Hellenic Syllogos presented at the time of the Treaty of San Stefano to Sir Henry Layard, the Greek or Hellenised population of the district of Perlepe was set down at 40,000 souls, either resident in the town or scattered over ninety Greek and forty-seven mixed villages. Notwithstanding the evident exaggeration of these figures, it is doubtless true that two years and a half ago, at the time those returns were framed, it was just possible, by classing as Hellenised all the Bulgarophones who adhered to the Orthodox Church, to outnumber the Bulgarians who looked to a national Church as the emblem of national existence. Since then thirty months have passed, Bulgaria from a mere name has become a living reality, and the mutilation which it underwent at the Congress of Berlin has served not to damp but to stimulate the national feeling among those left without the pale of freedom. In spite of the redoubled efforts of the Greek clergy, in spite of the support afforded to it by the Turkish authorities, in spite of its own precarious position, the Bulgarian national Church, deprived of many of its pastors, suspected by the Government, placed under the ban of excommunication by the Holy Orthodox Synod, has never ceased to rise on the tide of the national revival; and to-day the Greek Archbishop of Monastir is fain to confess that in the whole district of Perlepe there are only six villages, and in the town eighty houses, left—in all, barely a thousand souls—who still acknowledge his spiritual jurisdiction.

Yet the heavy curses with bell, book, and candle which his Eminence has fulminated against Perlepe do not appear to have impaired its prosperity. Placed in a sunny corner of one of the most fertile table-lands of European Turkey; surrounded by vineyards and tobacco-fields, to which it owes much of its wealth and reputation; sharing with Monastir the broad acres of the plain and its golden crops of barley, wheat, and Indian corn,—it derives also no little profit from its transit trade, as it lies at the mouth of the two mountain-passes through which the Salonica-Mitrovitza Railway taps the rich plateaux of western Macedonia. If the ruins of the royal castle, still called King Marco's Seat, on a rugged crag behind the town, testify that Perlepe has had a past, its bright well-built houses, its clean streets, its thriving population, are unequivocal witnesses not only to a certain measure of present wellbeing, but to a future of infallible prosperity. Where every household, even to the humblest, considers itself disgraced if it has not always its stock of provisions for at least six months in advance, there can be but little cause for grumbling. And in truth, the inhabitants of Perlepe are singularly free from that whining querulousness peculiar to the Christian populations of Turkey. Yet they suffered considerably during the war from the suspiciousness of the authorities; and the irritation of the Mussulmans often found vent, even to within the last few months, in outrages on their Christian neighbours, for which the latter could never obtain redress. But the appointment of a new Mutessarif, Abdul Rahman Effendi, has checked these outrages; and the stationing of military posts throughout a great part of the district, to protect the postal road between Monastir and the Salonica Railway, has of late preserved this region from the depredations of marauders. Abdul Rahman Effendi, who seems to be a pearl among Turkish officials, has even succeeded to some extent in reforming the *bekjis*, a class of petty tyrants, who rank among the chief curses of the country. The *bekjis* are properly guardians, *gardes-champêtres*, appointed by each village for purposes of rural police, for protecting the harvest from spoliation, superintending the gathering of the crops, keeping watch in the vineyards, and defending the peasants and their chattels from marauders. Of these a village is allowed to have as many as it likes and chooses to support. Unfortunately, in practice this excellent institution serves very different ends, and villagers are often compelled to have many more *bekjis* than they require, and of a very different stamp. By threats or violence every petty local magnate, even bands of brigands, force upon the villagers their own nominees, of course for the mere purpose of spying out the land, and compelling assistance in their deeds of oppression, corruption, and lawlessness. Thus a village finds itself sometimes saddled with eight or ten of these strange "guardians," each of whom levies upon it in kind a salary of twenty, thirty, and even fifty *liras* per annum —not for any services which he renders, but for the very tyranny which he exercises over the wretched peasantry under the commission of his real employers. The local influence of the latter in the *mejlisses,* and among at least the subordinate officials, makes it worse than useless for a village to seek redress, as complaints are sure to involve only aggravated persecution. Abdul Rahman Effendi has gone energetically into the question, and has succeeded by his firmness in ridding many villages of the Perlepe district from their ruffianly guardians. But vested interests are not to be overthrown in a day; and he has not yet been able to obtain the sanction of the Government to the radical change by which he proposes to extirpate the evils of the present system. I understood, however, at Monastir, that Ahmed Eyoub Pasha had given it his support, and proposed to apply it throughout the province. In lieu of the present *bekjis*, who are chiefly recruited from broken-down brigands or officials under a cloud, the authorities would supply to every village as many men from the regular army as they might require and be willing to lodge and support. When first told of the plan, I was inclined to fear lest the soldiers might misuse the power thus placed in their hands; but both Christians and Mussulmans are anxious for the change. Indeed, the mutual confidence and good relations of the Turkish regulars with the peasantry of Macedonia and Thessaly afford a pleasant contrast to the bitterness of sectarian animosity in those provinces. During a ten weeks' journey, throughout almost the whole course of which I was accompanied by an escort of Turkish soldiers, who were changed at every military depot, I had ample opportunity to watch their behaviour. To myself they were always attentive, invariably cheerful, anxious to be of use, and grateful for the smallest kindness; in their dealings with the peasantry they were considerate and good-tempered, and though I was careful to make inquiries, I never had but two complaints against them, and those of a most trifling nature; while their good-fellowship among themselves, and their readiness to lend one another a helping hand, were always pleasantly conspicuous.

But to return to Perlepe. If excommuuication has failed to bring down on the prosperous little city the fire and brimstone to which it should have been devoted, it has had one social result of a very peculiar kind. Among the superstitions which the Greek clergy has encouraged in the minds of the ignorant peasantry of the East, there are few more widely spread or more deeply rooted than the belief in a certain class of ghosts or evil spirits called *vrykolakas.* The Rev. Mr Tozer, in his 'Researches in the Highlands of Turkey,' gives an interesting account of the various forms which this belief assumes. "The idea concerning them [the *vrykolakas*] is, he says, that some persons come to life again after death, sleep in their tombs with their eyes wide open, and wander abroad by night, especially when the moon is shining brightly. . . . The more malignant form of these spirits nearly corresponds to the vampire, and is supposed to keep itself alive by sucking the blood of men." But the more common form is less noxious, as it contents itself with playing all kinds of mischievous tricks, and frightening the people without doing any further injury; or, according to another description, "appears as a real living man, who has the peculiarity of roaming by night as a dog over heaths, pastures, and even villages, killing with his touch horses, cows, sheep, swine, goats, and other animals in his passage, appropriating to himself their vital forces, by means of which he has the appearance of being in continual health and vigour." The principal cause which changes persons into *vrykolakas* after death, is excommunication; and doubtless, when excommunication was a penalty but rarely imposed, the fear of being converted after death into a *vrykolaka* added not a little to the terrors of the ban, while the dread of being revisited after death by the unclean spirit of the friend or relative who might still be tolerated during life even when excommunicate, operated as a spell to close every house against the unfortunate victim of the Church's wrath. Among the various means employed for exorcising the *vrykolaka*, the most orthodox was to obtain the reversal by absolution of the excommunication pronounced by the Church; and as the payment of heavy fees was always a condition *sine quâ non* of such absolution, it was of course strongly advocated by all ecclesiastical authorities. But there was another and more informal process, by which indeed peace was not restored to the afflicted soul, but immunity could at least be secured from the pernicious results of its nightly wanderings; and this process, as being cheaper, and from a selfish point of view equally effectual, enjoyed considerable popular favour. There were held to exist in certain towns families which were the offspring of the *vrykolakas,* and were empowered as such to treat in their name, accepting peace-offerings, and securing in return freedom from the intrusion of their luckless ancestors. Shunned though they were for their connection with the spirits, the powers of which they enjoyed the monopoly, and the mystery which surrounded the exercise of their arts, compelled a certain degree of reverential fear, while the gifts of their petitioners more than sufficed to compensate them for their social isolation. One of these curious colonies dwelt at Perlepe, and enjoyed until recently a high reputation for the potency of its relations with the spiritual world. But the rapid spread and wholesale scattering of excommunication consequent upon the Bulgarian schism has been fatal to its pretensions. Familiarity breeds contempt; and the good people of Perlepe have of late years been made so familiar with excommunication, that, from learning to brave its consequences, they have ended by laughing at them. Now that nine-tenths of the Christian population of the district have been placed under the ban, the belief in *vrykolakas* is reduced to a palpable absurdity. The good Perlepiote who a few years ago shuddered and crossed himself devoutly at the mere mention of a *vrykolaka*, now shrugs his shoulders and says: "Ah, well! I suppose we shall all be turned into *vrykolakas* some day. The more the merrier!" So the *vrykolaka* is exploded; and the unlucky families who claimed connection with him gradually give up their sackcloth and ashes, their dishevelled locks, their nocturnal spells, and all the rest of their stock-in-trade, and sink down from their supernatural estate to the dull level of common, labouring mortality.

The road from Perlepe to Graczko rises rapidly over the spurs of the Babouna mountains, and before plunging into the long and winding pass which leads down into the valley of the Vardar, a fine retrospect is obtained of the plain of Monastir, Perlepe itself lying immediately at the traveller's feet in a deep depression of the long and rugged chain which forms the eastern boundary of the lofty table-land; while far away to the south-west, beyond the level expanse of the plain, the gardens and domes of Monastir can be faintly discerned nestling under the shadow of the double peak of Mount Peristeri. From the summit of the pass, where a goodly detachment of troops, told off for patrol-duty along the postal road, were encamped on the hillside, we descended in sharp zigzags through a steep defile into the valley of one of the tributaries of the Czerna, the ancient Erigon, which, passing between Mount Nidjeh and the southern end of the Babouna chain, carries the waters of the Monastir plain into the Vardar. The scenery was lovely, the slopes of the wooded hills on either side being clothed with all the varied tints of autumn, while here and there the cliffs closed in upon the water-course, and scarcely left room for the road to wind beside the brawling torrent. But with the exception of the half-way *han*, we scarcely passed a human habitation during the whole of the long day's journey, till, about nine hours after leaving Perlepe, we reached the broad valley of the Vardar, and saw Graczko lying before us on the last spur of the mountain-range. Mr Tozer has identified

Graczko with the once important town of Stobi, which in Roman times was the meeting-point of the four great roads which intersected Macedonia. But its site is now occupied only by a squalid Bulgarian village. So without troubling ourselves to find the ford which leads to it across the Czerna, we made straight for the railway station, some two miles north of the village; and having dismissed my escort, I spent the night in wrestling with the fleas, which seemed to be the only guests besides myself in the miserable Bulgarian *han* which did duty for a railway hotel. The next morning the solitary train which creeps wearily each other day from Uskub down to Salonica, came puffing and panting up to the platform; and in spite of persistent dawdling at every station, and sundry little untoward events, such as running over an unfortunate cow and waiting half an hour for a Pasha's harem, we reached Salonica by eight o'clock in the evening, having only taken eleven hours and a half to cover a distance of a hundred miles. The next day I took the steamer for Volo, and on the following evening was back again at Larissa.

# CHAPTER VII.

## THE VALE OF TEMPE AND THE PLAIN OF THESSALY.

A SUCCESSION of severe storms which followed my return to Larissa warned me that, if I wished to carry into execution my plan of crossing the Pindus into Epirus, I should not tarry longer on the road, under penalty of being delayed, and perhaps driven back, by heavy snows and swollen torrents. My preparations were soon made—Hidayet Pasha, the Mushir in command of the troops in Thessaly and Epirus, and Khalil Bey, the Mutessarif of Larissa, vying with each other in making every arrangement which could conduce to the comfort and safety of my journey. The kindness and courtesy which they showed me, notwithstanding the bitter feeling which the policy of England towards Turkey was at that time calculated to excite—and in many cases, as I afterwards experienced, did excite—against Englishmen in general among the Turkish official classes, deserve to be specially acknowledged. Though belonging to different schools, both of which unfortunately have found but too few disciples among the officials of the Ottoman empire, both the Mushir and the Mutessarif were representatives of the best and rarest class of Turkish officials. Hidayet Pasha was an *old Turk*. Without powerful connections, without exceptional abilities, he had risen by sheer pluck, endurance, and honesty from the ranks to the highest military honours of the empire. He supplied his lack of education by sound commonsense; and though a fervent Moslem, his natural kindliness raised him above narrow sectarianism. Popular with the people, hospitable to his friends, and courteous to all men, he succeeded, by the simple exercise of that tact which is born not of diplomacy, but of the fulness of a generous heart, in gaining the respect and affection of a hostile population, who were inclined, both from political antipathy and from past experience, to look upon every Turkish official as the embodiment of tyranny. Khalil Bey was a *young Turk.* He had been attached for many years to the Ottoman Embassy in Vienna, and had visited most of the European Courts. He had spent those years profitably in gaining acquaintance not with the vices but with the virtues of Western civilisation, and he had returned to his own country imbued not with the hollow shibboleths but with the practical spirit of the nineteenth century. While residing in Austria, he had been able to study the means by which an empire composed of elements fully as heterogeneous as are to be found in Turkey can secure its unity without checking the individual development of its component parts. "Turkey," he once said to me, "is like unto a flock of sheep pent up in a stony field, which, finding nothing but rocks and weeds and brambles to feed upon, are continually trying to break through the fence in order to feed on the richer meadows of their neighbours; and as the shepherds in the adjoining meadows are always piping prettily to allure them over, while from their own shepherds they get only kicks and blows, the poor beasts will naturally never be quiet until they wriggle themselves through the holes in the fence. Our head shepherds in Constantinople fancy that by ordering from abroad the pictures of the fine animals which have gained prizes in the cattle-show, and by printing in broad-sheets the recipes according to which they are fattened, our own sheep must perforce grow fat and sleek and comfortable. Others imagine that if only a big wall could be built round our field to shut out the view of the bright meadows outside and the sound of the neighbours' *chalumeaux*, our sheep would soon make the best of their stones and weeds and brambles, and thrive upon them as if they were the sweetest grass. My own opinion is that neither the prize models nor the wall will prevent our sheep from either sickening and dying, or trying to make good their escape out of the fold. But if each of the shepherds in charge of our flocks would take the trouble to pull out the stones and the weeds and the brambles in his part of the field, and if the water which is wanted to fertilise it were not all carried off to feed the fountains and cascades of the head shepherds' pleasure-gardens, and if we only exchanged the stick for the *chalumeaux*, and gave our flocks a little more piping and less beating, our sheep would soon grow to like their fold, and we should be able to do without both the wall and the prize models." And certainly Khalil Bey, though he has charge of the most obstinate of all the flock, and though he knows that they may at any moment be transferred to a new owner, does what he can to temper to them the cold wind which blows over from Constantinople. The friendly, not to say cordial, relations which exist between the Greek Consulate and Archbishopric on the one hand, and the civil and military authorities of the province on the other, notwithstanding the delicate position of both parties, are a signal testimony to their mutual success in preserving the conduct of local affairs free from the baneful influence of political rancour. An instance which occurred just before my arrival at Larissa may be conveniently quoted as a case in point. The Military Governor of Epirus, a fanatical Circassian, of whom more anon, had caused two Greeks of Yanina to be arbitrarily arrested on the charge of seditious language, which was in reality the idle gossip of a coffee-house, and sent them in chains to Larissa under military escort, to be thence conveyed to a place of exile in Mesopotamia. Their relations telegraphed the facts of the case to Hidayet Pasha, and his Excellency on his own responsibility cancelled the orders of the authorities of Yanina, and released the prisoners on parole, to remain at Larissa until the matter was settled by reference to the Sublime Porte. No one could have condemned Hidayet had he washed his hands of the whole business, as he had, strictly speaking, no power to deal with it; but his acute sense of justice and of honesty dictated to him the more dangerous course of opposing his Epirote colleague, even at the cost of provoking the resentment of the Palace influences, which the Circassian Pasha is known to command.

The day before I left, I had an opportunity of witnessing in another form the harmony which reigned between the Greek and Turkish communities of Larissa. The Archbishop had in the morning given his niece away in marriage to one of the leading Greeks of the town, and in the evening a banquet took place at the Archiepiscopal Palace to celebrate the auspicious event. Priests and pashas, Mussulman beys and Greek merchants, responded to the invitation, and the merry temper of the guests, the gentle flirtations of Turkish officers and Greek ladies, the heavy fire of ponderous jokes, entirely blinded me to the horror of the situation, until the Mushir, in proposing the Archbishop's health, returned thanks on behalf of the Turkish wolf for being allowed to lie down once again with the Greek lamb; and I am sorry to say that his Eminence was himself so blind to this unholy alliance between the Crescent and the Cross, that he actually treated the allusion as a joke, saying he could discover no wolf in the room save the bridegroom, who had come to carry off his own lambkin,—upon which the fair lambs again turned their attention to the wolves, and looked as if they too would not mind being carried off in the same way; and a good deal of pleasant banter, not unaccompanied with whispering and blushing, went on in a true Christian spirit, till the signal was given to return to the *selamlik*, where wolves and lambs were stirred by the vigorous strains of a Turkish band to terpsichorean efforts, wherein, I am free to confess, they displayed more energy than skill. But its very novelty added zest to the exercise; and when I took French leave to retire at an early hour, in view of the morrow's start, I still saw the kindly old Archbishop standing in the middle of the room and beating time with foot and finger as the enthusiastic couples glided and stumbled around him.

My pleasantest recollection of Larissa will be the moment when, halting two miles outside the town on one of the tumular mounds studded about the plain, I turned to cast a farewell glance on the yellow patch of mud-houses and mud-streets, half veiled in the silver mist of malaria, which marked the capital of Thessaly. My course lay north-east across the plain towards the gorge where the Peneus, with an energy long since lost to its lazy waters, erst forced its passage to the sea between Ossa and Olympus. For to leave Thessaly without performing a pilgrimage to the Vale of Tempe would be a sin which not even the terrors of brigandage could excuse. After three hours' riding across the flat, the monotony of which is more than redeemed by the glorious prospect of the mountains clothed with the early lights of morning, the plain is broken by projecting spurs thrown out from the lofty slopes of Ossa. The Peneus disappears amid clumps of spreading plane-trees and silver willow, brawling streams come tumbling merrily down from the hills, the fields are fringed with hedges of blackberry and yellow thorn, and presently a tapering minaret, rising out of a dark grove of cypress-trees, marks the Turkish village of Baba, which lies at the western entrance to the gorge. Baba owes its name to a famous dervish, Baba Osman, who came into the country with the first Mussulman conquerors, and selected this favoured spot to found a *tekke*, or monastery. When Baba Osman died, wonderful miracles were wrought at his tomb: sultans endowed his *tekke*, and the dervishes of Baba became a power in the land. But now its prosperity has long been on the wane, though pious Moslems still journey from afar to the holy shrine, and the Christian peasants of the neighbourhood still hang their votive rags to its venerable cypress-trees, whether to conjure the Evil One or to propitiate Baba Osman's spirit, I was unable to discover. I had scarcely installed myself for my mid-day's rest in the quaint old graveyard which adjoins the *tekke*, when the old dervish who is now the sole inmate of the monastery sallied out to greet me. Now there are dervishes and dervishes. If in some of them the worst form of Mussulman fanaticism seems made incarnate, there are many of a more liberal, because perhaps more sceptical, school, who have tempered the harsh exclusive dogmatism of Sunnite orthodoxy with the milder inspirations of Buddhistic pantheism. When the grey-bearded patriarch of Baba spread out his hands as in prayer upon the stranger, and gave him, a Christian, the welcome peculiar to the true believer, "El salaam aleikoum," I knew that I had no fanatic to deal with. After the ordinary expressions of welcome and frugal show of hospitality, he insisted upon taking me himself round the *tekke* and into the small mosque where his first ancestors were buried; for dervishhood runs in families, and the eldest son of a dervish is invariably brought up to the same vocation. In the largest of the tombs, covered with a broad green mantle, rests the founder of the *tekke*, and to the wall are suspended his sword and his Koran. Many were the tales he told me of the wonders they had worked; but he added, *naïvely* enough, "God has now withdrawn His strength from them, and I am told I shall soon have to take them down from yonder wall, where they have so long hung, an object of veneration throughout the land, and wander forth with them in my old age to the far country whence we came." And when I asked him why he could not remain and finish his days in peace under Christian rule, he added: "To me all men are sons of God; but Baba Osman (the mercy of God be upon him!) lived in other days, and his sword is still red with the Giaours' blood. It would not be well that it should fall into their hands."

From Baba the direct road through the Vale of Tempe runs almost alongside the Peneus (which the Greeks now call the Salemvria and the Turks the Göstem); but the more interesting route runs up between terraced vineyards to a lofty plateau, where Ambelakia lies buried in a perfect bower of verdure. More than four hundred houses, many of them of stately construction, nestling amid groves of chestnuts and of plane-trees, still bear witness to its former prosperity, but two-thirds of them at least are now untenanted and fast falling into ruins. The twenty-four manufactories which at one time supplied the markets of Eastern Europe with famous cotton-yarns dyed with the rich red madder from Asia Minor, stand desolate and silent; the long caravans which used every year to convey some 5000 cwt. of yarn overland to Pesth and Belgrade have long since disappeared from its deserted streets; the high school, which once rivalled the best Greek colleges of Smyrna and Constantinople, has dwindled down to a mere village class-room where peasant children painfully spell out their A B C; the library, which ranked with those of the Holy Mountain, has been scattered to the four winds. English spinning-jennies first shook the commercial supremacy of Ambelakia; the Turkish soldiery completed its ruin during the Greek war of independence. From this melancholy abode of decayed prosperity another steep path leads down again into the valley. The gorge rapidly narrows. Olympus on one side, Ossa on the other, throw out their gigantic buttresses almost down to the water's edge, scarcely leaving room for the fringe of oleander and plane-trees intermingled with the darker green of olive and of oak, and the laurel sacred to Apollo, which overhang the placid waters of the Peneus. The ruined walls of an ancient castle rise to the left of the path; and far above, on a bold and lofty crag, a single arch stands out in sharp relief against the dark-blue sky. It is still called the Beauty's Tower; and a legend tells how a local siren, an Eastern Marguerite de Valois, used to cast the paramours whom she seduced into her stronghold down the precipice into the stream below. Here and there traces are visible of the Roman road which connected the plain of Thessaly with Thessalonica; and at one point a Latin inscription on a rock tablet records the fact that the Proconsul Lucius Cassius Longinus fortified Tempe. Of late the Turks have turned their attention to this road and repaired it with unwonted care—perhaps in view of a future retreat of their army along this route to Macedonia. The scenery presents that combination of the beautiful and the grand by which nature in her gentler moods tempers her awful majesty. The valley is a wild garden of broad-spreading trees and flowering bushes, *Tempe quæ silvæ cingunt superimpendentes;* while lofty cliffs of grey limestone, made bright with patches of red and yellow lichens, and with luxuriant vegetation growing out from every nook and cranny, tower above it to the right and to the left, now rising in one sheer unbroken wall, now broken up into a thousand fantastic pinnacles and buttresses. For about an hour and a half from Baba the road winds through the gorge, and then suddenly the valley expands. Olympus and Ossa fall away on either side in gentler slopes, and the Peneus rolls onward to the Ægean across a broad flat plain, where the yellow maize is still waiting for the reaper. The traveller to Salonica crosses the stream by a ferry near a ruined bridge, but our path keeps along the foot of the southern hills and between clumps of magnificent plane-trees and green ilex overgrown with a thousand creepers, and brings us in two hours to Tchai-Aghazy. It is a pretty little open roadstead which does a good deal of trade, and a good deal more smuggling, with small Greek craft from Salonica and the ports of Greece. Much of the produce of Northern Thessaly finds its outlet here, and the village, which is purely Greek, is prosperous. The Turkish Government has of late cherished the hope that if Larissa were left to Turkey, Tchai-Aghazy might be made to rival Volo. But were even Northern Thessaly to remain under Ottoman rule, Tchai-Aghazy could never be converted into a first-class port without an enormous outlay. Sheltered though it is from the south and south-east winds by a spur of Mount Ossa, it is open to the north-east gales which in winter sweep across the gulf from Salonica. The beach is shallow and anchorage is bad. Only two years ago a Turkish gunboat was driven on shore near here and went to pieces. Two small Greek coasting barques, that certainly did not run over fifty tons, were lying at the time fully half a mile from the shore. However, a ridge of rocks and the remains of an ancient mole run out in a north-easterly direction, and a good breakwater might no doubt be built there at no great expense, and insure a fair amount of safety to moderate-sized vessels. But should Thessaly ever enjoy its proper share of material prosperity and development, the future of Tchai-Aghazy as a summer resort and watering-place should be assured of success. A more lovely position can scarcely be imagined. Lying amid shady groves of trees under the forest-girt slopes of Ossa, which shelter it from the ardour of the south, it catches the fresh breezes across the gulf. As I stood on the beach in the evening, the sun had already sunk behind the massive chain of Mount Olympus, and its eastern slopes were clothed in the purple mists of evening; but its majestic domes were still bathed in sunlight; the glory of the heavens was reflected on the glassy waters of the gulf: to the north the faint outline of Mount Khortiatzi marked the bay where Salonica lies; and far away to the east the cone of Mount Athos, the Holy Mountain, strangely luminous and transparent, rose like a fairy vision out of the bosom of the Ægean Sea.

In a country where his time is limited, the traveller generally grudges having to retrace his steps along the same road. But the Vale of Tempe is too beautiful to allow of such regrets. The next day, on my way back, its grandeur revealed itself in a new shape. Storm-clouds had gathered about the mountains, the summits of the cliffs were lost amid dark lowering nimbi, and the voice of the thunder-god was heard muttering on Olympus. But the sun ever shines at Baba, and on issuing from the defile all was placid and serene. From Baba a road crosses the Salemvria and leads along the northern edge of the great plain of Thessaly to the large wine-growing village of Tirnowa. This is the district of Thessaly which was first occupied by Mussulman immigrants; for, before the conquest of the country by the Turks, the Greeks of Larissa applied to a Mussulman chieftain to protect them against the predatory incursions of the Bulgars, and in response to their invitation 5000 families from Asia Minor settled in the valley of Vereli at the mouth of the gorge of Tempe, and along the southern spurs of Mount Olympus, and thus formed a barrier against the northern invaders. In about three hours Tirnowa is reached, a large straggling village, or rather town, of about 6000 inhabitants, through which I had already passed on my way to Monastir. It lies pleasantly enough amid vineyards watered by the Europus, and its mixed population of Greeks, Wallachs, and Turks are unusually prosperous,—for besides its extensive wine trade, Tirnowa is one of the few places in the province which can still boast a native industry; its cotton prints and woollen tissues are to be seen in almost every homestead of Thessaly. From Tirnowa the road to Trikalla crosses the Europus and reaches the Salemvria near some ruins which are supposed to mark the site of the ancient Larissa. The modern town lies about six miles farther down the stream. Hence to Trikalla is a wearisome thirty miles' ride across the plain. At this season of the year, after the harvest has been gathered, the landscape is brown and bleak. The villages are squalid and ill-favoured.

The houses are mere hovels built of mud, often without even a tree to redeem their gracelessness. Yet if you go inside them you will find a measure of comfort which many an Irish peasant might well envy. There are bright pots and pans displayed along the whitewashed wall, a good rush-matting on the mud-floor; the wooden divan which lines one side of the room is covered with a gay bit of Elassona carpet or Tirnowa print; in that farther recess the mattresses and coverlets which constitute an Eastern bed are carefully stowed away against the night; and in the corner, where a smoky lamp is burning beneath the family *icon*, a baby wrapped in the tightest of swaddling-clothes lies peacefully asleep in a wooden cradle of many colours. The *chiftlik* of the Turkish Bey to whom the village belongs, though he may perhaps count his income by thousands, does not boast much greater luxury. True, a rickety wooden staircase leads up to a second floor, while the peasants' houses seldom possess an upper storey; his divan is covered with a tenth-rate Manchester print, instead of the more solid tissue of the country; and a clock which has long ceased to go, or a vase of paper flowers, shows the pretensions of the master; and there he lives among the peasants, whenever his presence is required on his estates, sharing their frugal fare and boorish ways. At this hour the village is deserted save for a few urchins playing about the well, and a few old grandmothers spinning or dozing in the doorways. All that can toil have long since turned out abroad, some to plough the fields, some to the threshing-floors where the golden corn-cobs of the Indian maize are waiting to be picked and sorted, others to the vineyards where the grape is just ripe and ready for the vintage. But we shall meet them farther on along the road: men in *fustanellas* and leggings, and thick cloaks of coarse grey homespun; women in dark-blue serge petticoats, and braided bodices fastened high up round the waist with big silver clasps; children in fragmentary nondescript garments, yet warm and comfortable withal; and every one well socked and shod, which is always in the East a sign of comparative affluence. A large herd of buffaloes, useful if ungainly animals, are wallowing in a marshy pool by the roadside to rid themselves of the flies, one of the chief plagues of Thessaly. Heavy waggons full of grain for Larissa, and curious barrel-shaped carts laden with grapes and drawn by sleek grey oxen, creak slowly along on ponderous wheels hewn out of the solid oak. Though the race of horses and of horsemen which once made Thessaly famous as the home of the Centaur has long since died out, there is still many a useful bit of imported horse-flesh to be seen about the country. Yet, with all these indications of material prosperity, it is painfully easy to see the moral havoc wrought by centuries of ignorance and bondage. The men are either cringing or surly; they are little better than serfs attached to the soil they till—for among the peasantry of the plain there are scarcely any who own the land upon which they live—and their bearing is that of serfs. The women, prematurely worn by hardships and exposure, have a hard degraded look: even here, among Christians, they are treated like mere beasts of burthen. Often have I seen them tramp along the road bent double under a heavy load, while husband or brother slouches along empty in front of them, or sits dangling his legs from the side of a waggon.

Misgovernment, indolence, and ignorance have not only cast their blight upon man—they have even marred the generosity of nature. A province which might easily maintain a million souls scarcely suffices to provide for 350,000 inhabitants. Owing to the enormous size of many of the estates—fifty, sixty, eighty thousand acres being often held in one hand—the landowners seldom feel the need of bringing the whole of their property under cultivation; and as the soil is light, and no artificial means are used to stimulate its productiveness, land is often allowed to lie fallow for two or three years at a time. Moreover, the amount of pasture-land is out of all proportion to the grain-producing area. Thus it happens that of this rich plain of Thessaly not more than one-fourth or fifth is actually under cultivation. Yet in good years Thessaly has yielded 1,000,000 Stamboul *kilehs* of barley (the Stamboul *kileh* is rather less than half a hundredweight), 1,800,000 *kilehs* of wheat, 1,200,000 *kilehs* of Indian corn, 3,000,000 lb. of tobacco, besides other smaller crops of rye, oats, beans, millet, &c.; and even these figures multiplied by four or five would be far from reaching the possible yield of this enormous garden, were as much ingenuity applied to the development of its resources as the Turkish Government display to paralyse them.

But statistics and considerations on the wealth and possible yield of brown fields and fallow land do not suffice to relieve the monotony of a six hours' ride across the plain when a grey sky encompasses the landscape on all sides, and impresses it with its uniform dulness. Right welcome, therefore, was the view when, towards evening, the sun, just sinking behind the Pindus, broke through the clouds, and lit up with its last rays the minarets and ancient stronghold of Trikalla. Trikalla, "The Thrice-lovely," does not perhaps quite deserve so ambitious an appellation, but its position is certainly picturesque. Its straggling houses, interspersed with trees, spread up the slopes of an isolated hill at the end of a long low ridge which the Kambunian chain throws out into the plain of Thessaly; and, rising above the town, an old medieval fortress, still jealously held by a Turkish garrison, forms an ornamental, if no longer useful, feature in the landscape. A quaint square clock-tower bears witness to the rule of Latin princes in the land, and, like all similar constructions, is popularly ascribed to the Genoese; but it has long since been taught by the conqueror to toll out the hours *à la Turque.* Trikalla is a sleepy town of about 8000 inhabitants, mostly Greeks and Wallachs, with a few wealthy Moslem families and a small colony of Jews. Its position is naturally strong, and of considerable strategic importance, as it commands the *débouché* from Epirus down the upper valley of the Salemvria. In Hellenic times Tricca was a famous seat of learning, sacred to Æsculapius, and its medical university was the resort of aspiring M.D.'s from all parts of Hellas. But nowadays all that is forgotten, and the Trikalliotes are reputed for anything but intelligence or instruction. Greek schools are doing something towards rousing them from their coma, but the intellectual standard is still very low even for Thessaly. Mine host is a rich landowner and merchant, the first Greek notable of the town, and his income, he informs me, not without pride, exceeds

£3000 in good years; but wealthy and worthy as he may be, his mind appears as scantily furnished as his wardrobe. His political opinions, if simple, are, however, at least robust and commendable: "Confusion to the Turks, and long life to Gladstone! only it's a pity that he should have been born a Bulgar."

# CHAPTER VIII.

## THROUGH THE HEART OF THE PINDUS.

On the evening before I left Trikalla, the Turkish Governor insisted that I should swell my escort, which had hitherto consisted of ten *suwaris* or horsemen, by taking with me a small detachment of infantry. Now doubtless a numerous escort enhances one's importance, commands respect, and adds to the picturesqueness of one's cavalcade; but it is also apt to impede one's progress, and materially increases the expenses of the journey, and I strongly resisted, though in vain, the favour thrust upon me. But the first piece of news I heard in the morning was well calculated to dispel any lingering hesitation. It was supposed that, thanks to the energetic measures of the present Mushir, brigandage had been wellnigh stamped out of the plains of Thessaly; and in truth there had been of late but few cases of highway violence in the province, and those only in the mountain districts. Anyhow, the brigands were out again, and no mistake about it. On the preceding Tuesday they had seized, near Armyro, two Mussulman farmers and a petty Government official; and on the following Thursday they had waylaid a wealthy Turkish gentleman, Arif Bey, the President of the Municipal Council of Salonica, who was on his way to visit his farm at Velestin, on the highroad between Larissa and Volo, and after killing two of his escort, they had carried him off to the mountains, and demanded 9000 *liras* ransom. After this piece of intelligence, when my escort arrived, I counted the thirty men, cavalry and infantry, and found them not one too many.

From Trikalla the valley of the Salemvria makes a sharp bend to the north-west, the stream descending from its mountain home in the Pindus between the precipitous slopes of Mount Kotsiaka (the Eastern Pindus) and the low spurs of the Kambouni, at the extremity of which lies Trikalla. To the south the Agrapha, or

Mountains of the Unwritten Villages—so called from the privileges granted by the Ottoman sultans to their free Wallach populations—stretch far away into the kingdom of Greece, a mass of peaks and crests of exquisitely varied forms. But the ridge which rises to our left is a lofty unbroken cliff, averaging 4000 to 5000 feet in height, a gigantic natural wall twenty miles in length, dividing Epirus off from Thessaly. Opposite to the northern extremity of this wall, on the left bank of the Salemvria, rise the strange columnar rocks upon which are perched the famous aerial monasteries of Meteora. Seldom does nature show herself more lavish of rich colours and fantastic shapes. Masses of conglomerate, cleft asunder by some primordial cataclysm, have been chiselled by the hand of time into the strangest forms of columns, pinnacles, pilasters, bastions, towering above the valley. The deep ravines which intersect them are clothed with the most luxurious vegetation, while rain and sunshine have painted their grey cliffs with rich streaks of yellow, brown, and madder-red. The ascetic fervour of the early Christian ages scaled these inaccessible heights; but many centuries elapsed before the monasteries now perched aloft were built. The first hermits of the Meteora doubtless dwelt in the rock-caves which still honeycomb its rocks. It was only in the fourteenth century, at the time when the Servian so-called Paleologos Simeon Orosch reigned in Thessaly and Southern Albania (1367), that the monk Nilos obtained permission from Bessarion, Bishop of Staqus, to found four churches on the rocks of the Meteora, and thus laid the foundation of the monkish republic, which emperors afterwards endowed and visited. Since then the Meteora rocks never ceased to be a favourite retreat of Eastern monks, until the confiscation of their property in Wallachia of late precipitated their decay. To-day many of these holy dwellings are tenantless, while others are only occupied by two or three inmates. Altogether, there are only seven monasteries now inhabited out of twenty-four; and the pious colony, which used to number from 500 to 600, has dwindled down to 21. No new recruit has arrived at Meteora within the last twenty years; and when the present generation has died out, the traveller will be condemned to stand at the foot of the rocks and look up from afar with vainly curious eyes at these strange monuments of a time-expired piety — for the secret or the courage which enabled the pioneers of Christian asceticism to scale those walls has long been lost. Nowadays the monks, true fishers of men, let down a net from their lofty perch, and by means of a rope and windlass haul the visitor up to their quaint mid-air abodes. The Monastery of St Stephen, and one or two smaller ones, are alone approached by a drawbridge thrown over a deep cleft in the rocks. As my time was limited, I was only able to visit one of the monasteries, and my choice fell upon that which is called, *par excellence*, the Great Monastery of Meteora. Like all the more important monasteries of this group, it was withdrawn at an early date from the jurisdiction of the local bishops, and placed by special indulgence immediately under that of the Patriarchate at Constantinople. Although second in size to St Stephen's, it boasts the finest church, and a rich treasure of ancient books and manuscripts. The clatter of our horses' hoofs up the ravine over which it towers, brought one of the monks to the overhanging balcony which forms the only entrance to the monastery. It is not every comer who is admitted to these eyries, especially in the present troublous times; for they have been for centuries the savings-banks of the peasantry of Thessaly, who intrust their hoards to the safe-keeping of these monkish strongholds. In reply to our shouts for admission a small net was lowered, into which I put the letters of recommendation which I had obtained from the Metropolitan of Larissa. Their contents having been found satisfactory, a servant was lowered in the larger net which is used for living freight; and having taken his place, I presently found myself hoisted in mid-air, cramped up in the meshes of the net, and feeling altogether uncomfortably helpless. Three mortal minutes does this aerial journey last; and it was with a sense of pleasurable relief that I felt the net being caught by a long hooked pole and dragged on to *terra firma*. I was speedily released from my cage, and right hearty was the welcome which the old monks gave me. The monastery is composed of several rickety, rambling wooden buildings—

built, however, on solid foundations of stone. In the centre rises the church, a small but handsome Byzantine structure, the inner walls still rich with ancient frescoes. But the rows of carved wood stalls are nowadays scantily tenanted. There are only four occupants left in the monastery, which once counted over a hundred inmates. In olden days every monk was taught a trade, so that the monastery was able to supply all its own wants. But the workshops are now empty. The library, with its fine collection of parchments and vellum-bound volumes, is deserted, and the dust is allowed to accumulate undisturbed on its shelves. The youngest of the four monks is over sixty; and when the last one dies, the solitary servant of the monastery will climb down the face of the cliff on the giddy ladder which forms the only other means of communication with the world below, and the Great Monastery of Meteora will be abandoned to the havoc of the elements, until there remain of it but a name, *ut pueris declamatio fiat.* Again I ensconced myself in my cage and was pushed into space. The descent was certainly more rapid and pleasant than the ascent; and after waving a farewell to the good monks, who were watching me from their weird abode, 280 feet above the ravine, we returned for the night to Kalabaka, a large village which lies at the foot of the Meteora group of rocks.

The Archbishop of Larissa had furnished me with a letter of introduction to the Bishop of Stagos (Kalabaka), and his Holiness kindly insisted on my accepting his hospitality for the night. Nor was I inclined to resist the invitation; for, apart from the promise of good cheer and clean quarters which the episcopal residence held forth in marked contrast to the squalid hovels of the village, I was anxious to make the acquaintance of a man whom I had heard mentioned at Larissa as of original character and superior abilities, and whose appointment to the bishopric of Kalabaka was looked upon as a sentence of semi-exile which he had incurred by his heterodox views in matters both spiritual and temporal. He was a man of handsome and somewhat haughty presence, still comparatively young, with a striking countenance, eyes deep sunk and defiant under heavy eyebrows and a massive forehead, a long well-shaped nose, with delicate nervous nostrils, and a strange sarcastic smile on his thin, straight lips, scarcely disguised even under the heavy moustache and flowing beard. Almost more striking at first sight was the evident care bestowed on his dress and person. The Greek clergy, even in the highest ranks of the hierarchy, are generally slovenly in their personal appearance. The Bishop of Kalabaka was a marked exception to this rule. Buried for years in an out-of-the-way village where his only associates could be boorish peasants, he had lost none of the refinement which with him was evidently the habit of a life. The silk cassock was in keeping with the delicate white hands and the perfumed beard and hair, while the bold head and erect figure redeemed the outer man from effeminacy. The inner man was more difficult to gauge. His conversation was exceedingly reserved, and he circumvented all leading questions with a truly oriental suppleness. Only here and there an occasional word, a sudden light in the eyes, a silent smile, gave the measure of the bitterness and mocking doubt to which methinks the hard lips would often fain have given open expression. The dreamy, useless life of the ascetes who had fled from the temptations of the world to the safer solitude of the Meteora rocks, was not likely to meet with much sympathy from a man who was evidently born to brave the dangers of the struggle. Nor did he care to conceal his contempt for the pious drones. When I confessed to a feeling of regret at the doom which threatened to overtake the once flourishing colony of hermits, he merely shrugged his shoulders: "Vanity of vanities, all is but vanity! and even a vanity which has lived for 800 years scarcely deserves a funeral oration." And when I pressed him further—"Do you recollect," he added, "the parable of the talents, and the servant who buried his in the earth? You should have asked the *Hegoumenos* of the Great Meteora to read it to you?" The point was clear and incisive, and required no further comment.

When, later on, I stretched myself out to sleep between the silken sheets with which the luxurious forethought of mine host had furnished my bed, I fell to wonder how this man of delicate tastes and refined intelligence contrived to fructify the talent which to him too had been intrusted, amid surroundings so rude and uncongenial to his temper. He seemed in the morning to have divined my thoughts. "Your horse is not yet saddled, and the pack-animals are not yet ready; would you like to come round and visit my school?" A large barn adjoining the Bishop's house had been turned into a schoolroom. A youth who had studied at Yanina and Athens acted as a school-master (he had been sent for, I afterwards understood, by the Bishop, who paid his salary out of his own pocket); and as we entered, he was pointing out a lesson of geography on an excellent map of Kiepert, which covered one of the walls of the room. The children, some two dozen or more, were dirty and ragged, peasants' children, who had to work in the fields during six months of the year; yet there were many bright and intelligent faces among them, and they answered the questions which were put to them promptly and correctly. Some of them then read and spelt for their visitor's benefit, and did some simple sums on the board with accuracy and readiness. With the children, the

Bishop's pride seemed to unbend, his smile lost its bitterness, and as he patted them on the head, their bright, frank look met his eyes with affectionate fearlessness. As we left the room, I wished him all success for himself and for his school; "You at least," I said, "do not mean to bury your talent in the earth." "Bah!" he answered, with a hard, bitter laugh, "my talent has been buried long ago: I did not bury it myself, but others took care to bury it for me. Yet perhaps, among those children, some whom I have equipped for the battle of life may carry forth their talent into the world and make it fructify tenfold."

From Kalabaka it is a long and hard day's journey to Metzovo. The first part of the route lies up the valley of the Salemvria, between shady groves of plane and maple trees, of which I have never seen more magnificent specimens than in Thessaly. On either side the hills are clothed with timber, while here and there Wallach villages peep out of the dense foliage. But brigandage desolates this fair country. Two villages under which we passed had been within the last ten days sacked and partly burned by bands of dastardly marauders. On the way we meet numerous bodies of the unfortunate peasantry abandoning their homes and flying for safety to the plains; farther on the Wallachs driving their herds and flocks down to their winter quarters in the lowlands,—picturesque caravans of men, women, and children, and beasts of every kind, their small household goods packed on nimble ponies, with here and there a baby's head peeping out of a heap of wraps and blankets: a different type, too, from the Wallachs whom I had seen about Olympus—smaller, darker, and better featured, far more nearly approaching the Greek type than the so-called Greek populations of the plain. Five hours from Kalabaka, the Salemvria, which is no longer the placid, lazy stream of the plains, divides into three branches. Our route lies up the central valley, past the military post of Kalamash. The ascent becomes steep and rugged; the character of the vegetation changes—we leave the oak, the plane, the maple beneath us, and pass into dense forests of pine and naked beech trees, through which the cold mountain wind whistles and moans. As we rise the view expands; beyond the lower heights of the Eastern Pindus, the plains of Thessaly reach away to the long low ridge of Pelion and the pointed peak of Ossa. A dark depression marks the Vale of Tempe, and over it tower the snow-capped domes of Olympus; while to the north, behind the Kambouni chain, the mountains of Macedonia rise in jumbled masses, fading away into blue space. A little higher still, and everything is enveloped in rolling masses of grey mist; and in another hour, four hours from Kalamash, we stand on the watershed of Epirus and Thessaly, on the summit of the Zygos, a pass which culminates in bare rocks, forming a majestic gateway between the two provinces, 5640 feet above the sea. Heavy rains have fallen, and the rushing of many waters is heard on all sides, for this is the home of many mighty streams. From these heights the Arathus runs down to the Gulf of Arta; the Achelous, or Aspropotamos, flows across the Greek frontier to the mouth of the Corinthian Gulf; the Veneticos goes to swell the waters of the Vistritza; and the Salemvria descends to fertilise the lowlands of Thessaly. Dark clouds hang about the mountains of Epirus and shut in the view; but beneath us we look down into the deep valley of the Arta Metzovitico. The sides of the mountains are bare and bleak; the path, which is called a road, leads downwards in sharp zigzags, torn here and there by landslips and by foaming torrents: thunder and flashes of forked lightning harmonise with the wild grandeur of the scene. Over on the opposite side of the valley Metzovo hangs on to the precipitous rocks. It seems but a stone's-throw to its dark-grey houses, yet after two hours' toilsome descent we find ourselves only at the bridge which spans a northern branch of the Arta. Night has already closed in upon us, dark and gloomy, and we still have to climb up the other side of the ravine, picking our way among the rocks, the horses stumbling, the escort cursing, until at last, after twelve hours' travelling, we reach the welcome shelter of the *han* of Metzovo. A hostelry in Turkey consists only of a roof and a floor, already tenanted, perhaps, by unwelcome guests; but it affords food and rest, and neither men nor horses are inclined to cavil at the quarters which supply these two desiderata.

# CHAPTER IX.

## AMONG THE WALLACHS.

It is almost a miracle how Metzovo holds on to the precipice against which it is built. Its square, grey stone houses rise tier upon tier, clinging as best they may to the rocks upon which they are perched; above them a mountain-wall 2000 feet high—beneath them a deep ravine, where, 1000 feet below the town, three brawling torrents join to form the Arta Metzovitico. On the opposite side of the gorge a cluster of houses, under the shadow of towering cliffs, form the suburb where the sun never shines—Metzovo Anhelion. The position of Metzovo must have been at all times one of surpassing strategical importance, commanding as it does the only practicable pass between Thessaly and Epirus; and it is probably to this cause that the very existence of the town amid such inhospitable precipices is due. The picturesque battlements of the castle, which still overshadow its houses, no doubt mark the site of a far more ancient citadel. The chief interest attaching to Metzovo arises, however, from its being the most important centre of the Wallach race. Not only does it still enjoy a certain ecclesiastical independence—as the Exarch is not placed under the jurisdiction of the Metropolitan of Yanina, and is directly responsible to the Patriarch of Constantinople—but it was for a considerable time endowed with many valuable political privileges, which were conferred upon it by a Grand Vizier of the seventeenth century, who took refuge at Metzovo during a period of disgrace, and was mindful of the hospitality shown to him by the Wallachs in adversity when he was at last restored to imperial favour. Though these privileges were withdrawn thirty years ago by the Sultan after the insurrection in the Pindus, their effect has not yet entirely passed away. Preserved on the one hand from the Hellenising influence of the Greek clergy, and on the other from the immediate presence of Turkish misrule, the inhabitants of Metzovo, whatever their Hellenic sympathies may be, have maintained their Wallach character; and in its six or seven hundred houses there is not one where the Wallach tongue is not spoken, where Wallach traditions are not treasured up, where the old *civis Romanus* feeling is not still alive —"We are Wallachs, *Romounoi*."

The origin of the Wallachs is a question which has given rise of late to such angry polemic that it may be worth while to recapitulate briefly what is known of their past history.

More than four centuries ago the Byzantine Khalkskondylas, referring to the Wallachs of the Pindus, wrote: "They speak the same language as the Dacians" (now the Roumanians)—"a language which resembles that of Italy, yet so corrupt, and differing in so many points, that the inhabitants of Italy scarcely understand what they say: but whence these people, who speak the Roman tongue and have Roman customs, came when they settled down in these regions, I have never ascertained from any mortal, nor have I heard that any one has treated of the subject." Since that time the question has indeed been treated by many learned authorities, and, in the last few years, since the modern development of the so-called principle of nationalities, with all the passion and acerbity belonging to political discussions; yet in the main the results have been so contradictory that the question has lost but little of the obscurity which attached to it in the days of the Byzantine chronicler.

It was not till the latter part of the last century that the question attracted attention in Europe. The first effort to elucidate it was made by Dr Thunmann, a Professor at the University of Halle, in an essay "On the History and Language of the Albanians and Wallachs." The materials for this work appear to have been supplied to him by a young Macedonian from Mosehopolis, who proceeded to Halle to complete his studies; and besides presenting the Professor with a small lexicon of the Macedo-Wallach and Albanian languages, which had been published at Venice in 1770 by the Prototope of his native town, was able to furnish him with much practical information concerning the habits and whereabouts of those hitherto almost unknown races. With regard to the Wallachs south of the Danube, Thunmann arrived at the conclusion that they represented the remnants of the ancient Thracians, who, after being Romanised by intermingling with the Roman colonists and adopting the Roman tongue, were destroyed out of the town and plains by the successive barbarian invasions, and compelled to seek refuge in the mountainous regions of the Peninsula, the Balkans, the Rhodope, Mount Olympus, and the Pindus, where they maintained a wild and nomadic existence, surrounded on all sides by the new-comers, Slavs, Petcheneques, and Greeks. They were thus closely related to the Roumanians north of the Danube, who had also sprung from the intermingling of the Roman colonists with the old Thracian populations known under the name of Dacians and Getes, and who were also forced by the barbarians to fly to the mountains in order to escape destruction; while many of them, emigrating westwards, occupied Siebenbürgen, and all that part of Hungary lying north of the Danube, of which the

Hungarians afterwards slowly dispossessed them.

This theory was not slow in eliciting counter-theories; and already, in 1781, Sulzer published his History of Transalpine Dacia, in which he sought to refute the notion that any large body of descendants from the Roman colonists of Dacia could have survived the barbarian invasions north of the Danube; and as he failed to find any traces of the existence of any such nation up to the eleventh century, he ascribes their appearance after that date to emigrations from the Balkan peninsula, whence they brought with them to Catholic Hungary and Saxony not only the Roman tongue, but also the orthodox faith of their Macedonian brothers. This theory was further supported by Engel in his 'Commentatio de expeditionibus Trajani ad Danubium,' who accounts for the reappearance of the Roman tongue in Wallachia and Siebenbürgen by the drafts of Wallach colonists whom Krumus, a Bulgarian king, who ruled in those regions, brought back with him in 814 from a victorious expedition south of the Danube.

In more recent years this theory has been

questioned by Thomaschek, who does not consider the identity of language between the Wallachs of Central Thrace and the Roumanians north of the Danube sufficiently conclusive of an identical origin, arguing that the same Romanising influences working on kindred races, however territorially removed from each other, must produce cognate results upon their language and customs. But in Rösler's 'Romänische Studien,' the theory of the migration of the Roumanians found its fullest development. Dismissing contemptuously all the old arguments in support of the presence of a Romanic element north of the Danube before the thirteenth century, he peoples the Balkan peninsula up to that date with the Romanised descendants of the early Thracian population, claims for them a large share in the revival of the second Bulgarian kingdom, and then takes hold of the Mongol and Turkish invasions to make them migrate in a body to the northern banks of the Danube.

Though Rösler's theory was soon attacked on various points, it was not till last year that a Slav writer, Dr Pic, undertook to assail it all along the line. His object is clearly explained in the short preface to his work, 'The origin of the Roumanians,' where he says: "An attempt having been made to rob the Slav race of a not inconsiderable portion of its past, both in Hungary and on the Balkan peninsula, it is time that the voice of a Slav should be raised to vindicate the historical inheritance of his race." The strong bias of the author betrays itself even under a tone of almost exaggerated moderation; but if it robs the work as a whole of some of its value, it cannot diminish the strength of many of the arguments which he adduces. If he fails to make good his contention that Moldavia, Wallachia, and a great portion of Eastern Hungary, were throughout the dark ages of history, from the fall of the Roman empire to the beginning of the thirteenth century, tenanted by a sedentary Roman or Romanised population, he has, I think, at least succeeded in reducing to its proper value the preponderance claimed by Rösler for the Wallachs during the same period in the Balkan peninsula.

It does not come within the scope of these

jottings to give a detailed analysis of M. Pic's work; but I may be able with its help, and with that of the other authorities above referred to, to construct a rapid sketch of the history of the Wallachs south of the Danube.

It must be recollected that the number of Roman colonies established in the Balkan peninsula was not considerable; and though through them Latin obtained as the official language a footing in the country, it may well be doubted whether their influence was sufficient to Romanise the whole populations of Thrace. There are, in fact, many proofs to the contrary: the Greek colonies along the shores of the Ægean preserved their nationality throughout the days of Roman rule; the survival of the old Illyrian tongue in Albania shows that in that part of the peninsula Roman influences were never paramount; the fact that many Thracian tribes preserved their national organisation under their own rulers up to the third century also tends to prove that Thrace was never completely Romanised. On the other hand, on the northern bank of the Danube, in the province of Dacia Trajana, the California of the ancients, which attracted colonists from all parts of the empire, the Roman element must have rapidly acquired a complete preponderancy, and absorbed the native population. But this wealthy province was the first to fall under the weight of barbarian invasions. In 271 the Emperor Aurelian was obliged formally to cede it to the Goths, after carrying the greater part of the population across the Danube into the southern province of Dacia Riparia. The sacrifice, however, only stemmed the torrent for a while. One wave pushed the other forward. The Huns followed the Goths across the Danube, and devastated the whole of the peninsula; and when Attila's empire was broken up, the Huns only disappeared to make room for the Slavs and the Bulgars. Towns and villages disappeared; the fields were deserted; whole nations were destroyed. During this period of confusion, anarchy, and desolation, what became of the Roman colonies and the Romanised Thracians of the peninsula? Facts must here give way to conjecture; and the most unreasonable conjecture is certainly not that, flying before the storms which broke over the peninsula from the north and the north-east, they sought refuge in the wildest parts of the mountain-ranges which intersect these regions; and that the Wallachs, who appear in the pages of history about the beginning of the tenth century, are the descendants of these fugitives, who exchanged under dire necessity the culture and refinement of town life for the hardships of a nomadic existence, preserving only through the vicissitudes of fortune the language which they inherited from their ancestors.

The first allusion to the Wallachs is generally considered to be found in an episode of the Byzantine expedition against the Avars in 579, related by Theophanes. A soldier, seeing a pack-animal drop its burden, turned to a muleteer, saying, "*Torna, torna, fratre!*" The muleteer did not understand him, but some of the soldiers did, and fancying the enemy was upon them, cried out "*Torna, torna!*" and fled, creating a panic throughout the army. But when it is recollected how many mercenaries from all parts of the world served in the Byzantine armies, it surely does not absolutely follow that the sentence in question was Wallach; nor, even if that were proved, would it show that the population of the peninsula was at that time Wallach, —the less so that the muleteer, probably a man of the country, to whom the sentence was addressed, did not even understand it. The first authentic mention of the Wallachs occurs in 976, where a colony of them between Kastoria and Prespa, under the Kambouni, is alluded to in a Golden Bull of the Emperor Basil II. In the course of the eleventh and twelfth centuries other colonies of Wallachs are mentioned in the Rhodope, under the Balkans, in Thrace, in Ætolia, and in Epirus. In Thessaly, which, together with the adjoining districts of Epirus and Macedonia, was called from the twelfth to the fifteenth century Great Wallachia, the Wallachs formed the mass of the population; and it was there that Benjamin of Tudela met them in the course of his travels. But both his description of them, and the information to be gathered from other medieval sources, show conclusively that their colonies were the mere temporary settlements of nomadic tribes, under complete subjection to the ruling races, Greek or Slav, and possessing no national cohesion. So conclusive, indeed, is the evidence on this point, that it would scarcely have occurred to any one to claim any political influence for the Wallachs of that period, had not the Chronicles of Niketas Khoniates and the historians of the Crusades appeared to give the Wallachs, by the free, if erratic, introduction of their name, an almost preponderating share in the revival of the second Bulgarian kingdom, and its struggles against the Byzantines.

The Byzantine empire had succeeded, during the eleventh and twelfth centuries, in reducing once more the Bulgarians to subjection. The Emperor Basil II., surnamed Bulgaroktonos, after thirty years' fighting, broke up the powerful kingdom founded by the Czar Simeon, and Thrace was once more restored to the Byzantine empire. But the peace which was maintained by the strong hand of the Comnenes up to the fall of the last of that dynasty, failed to restore permanent strength to the rotten fabric of the empire. No sooner had the Emperor Andronicus lost his throne (1185) than a vast insurrection broke out, under the leadership of the two brothers Peter and Asen, throughout the northern and central portions of the peninsula, and resulted, after the vicissitudes of a ten years' struggle, in the erection of the so-called Wlacho-Bulgarian kingdom. Now the Byzantine writers persistently speak of Peter and Asen as Wallachs, and ascribe to the Wallachs not only the initiative of, and the chief part in, the insurrection, but the whole credit of its successful issue. Had we no other sources but these writers to turn to for our information, it might still be possible to question their veracity, on the plea, *fecit cui prodest.* The Bulgarians, who monopolised the fruits of the victory, must surely have borne a part in the battle; and had the Wallachs really taken the preponderating part ascribed to them in the war of independence, they would assuredly not have vanished again, immediately after its conclusion, from the pages of history. But there is no need for mere conjectures. Slav sources afford abundant proof that Peter and Asen were not Wallachs, but Bulgarians, and, as is mentioned in the letter of Pope Innocent III. to the King of Hungary, descendants from the old stock of Bulgarian kings. Nor did they lay claim, as has been maintained, to a Roman descent; but when the Pope, from whom they were seeking to obtain the recognition of their title, wrote to them that he had heard they descended from the noblest blood of Rome, they simply seized the cue which was thus given them, and with diplomatic dexterity thanked his Holiness for reminding them "of the blood and the fatherland from which we are descended." The title which these Bulgarian Czars assumed—*Imperator Bulgarorum and Wlachorum*—would perhaps alone show that they considered themselves rulers over the Bulgars in the first place, and in the second place only over the Wallachs. From all this it would appear that the Byzantines of the thirteenth century, in refusing to call the Bulgarians by their name until the triumph of the latter rendered all further prevarication useless, simply set the example followed by the Greeks of the nineteenth century, who also refused to recognise the existence of the Bulgarians as such, and continued persistently to call them Bulgarophone Hellenes, until the Bulgarian schism and the independence of Bulgaria exploded their idle quibbles.

That the Wallachs took a share in the struggle of the Bulgarians against the Byzantine empire, need not be gainsaid. But their share was not such as to insure them, after the victory, the enjoyment of equal, far less of preponderating, rights in the new kingdom. They still remain a subject race—pent up ever more closely in their mountain homes, until at last the Turkish invasion swamps both victors and vanquished, and again envelops in darkness the history of the peninsula. When they reappear once more on the surface out of the ferment of recent political events, their colonies in the Balkans and the Rhodope have disappeared—absorbed, probably, by the surrounding Bulgarian element. Hellenisation has also done its work in what was once Great Wallachia. The populations of Thessaly and Epirus have become, to all practical intents, Greek—Hellenised, no doubt, chiefly by the influence of the Greek clergy.

But in the wild mountain-ranges which separate Epirus, Thessaly, and Macedonia, and especially in the Pindus—which seems to have been the chief stronghold of their race—they have outlived the storms of centuries,—a compact population, with their own tongue and their own traditions, treasuring up the past until the hour of awakening has rung for each and all of the subject races of Turkey. It is not necessary, in order to claim for them a share in the future of the peninsula, to exaggerate the part they have played in its past history. No one who carries his eye along the map from Kalarrytæ to Metzovo and Samarina in the Pindus, down the Macedonian slopes of the Kambouni to Grevena, Kosana, and Servicen, and across the confines of Thessaly to Wlacho Livada and Elassona under the brow of Mount Olympus, can suppose that a hardy and intelligent race holding such commanding positions can fail to make its weight felt in the settlement of the destinies of these countries. And those are only the bulwarks of the Wallach race: they have their outposts in all the neighbouring towns of the three provinces; and every winter, when the snow lies deep on their mountain homes, they sweep down upon the plains, dotting them about with tents and sheds, and covering the pasture-lands with herds and flocks innumerable. Adventurous and enterprising, they present a rare combination of pastoral virtues and commercial instinct with a contemptuous repugnance for all agricultural pursuits. Almost the whole pastoral wealth of the country is in their hands; among the Albanians they are known only as the "Tchoban," or shepherds. Yet they show an equal aptitude for all commercial and industrial pursuits. Every year scores of young men leave their homes, to return only when their fortunes are made. The cotton and woollen tissues of Western Turkey, the coarse grey cloaks of the Greek and Albanian peasantry, the gorgeous gold and silver embroidery and inlaid weapons of which the Skipetar and the Palikar are so proud, the delicate wood-work which adorns the ceilings and panels of so many Albanian houses, are all the work of clever Wallach hands. But after years of toiling in the towns, the Wallach returns to his mountain home to enjoy the evening of life among his own kinsmen. In some respects there is no race among whom the national feeling is so strong as among the Wallachs. This national feeling has scarcely yet grown into a political movement. Hitherto the influence of the Greek clergy has prevented them from looking elsewhere than to Greece for the fulfilment of their destinies. But their faith has always sat light upon them. When the hand of the Byzantine emperors was lifted against them, they went over in a body to the Latin Church. When Latin emperors reigned at Constantinople, the spirit of opposition drove the Wallachs back to the Orthodox creed. They have since then adhered to it, because the Orthodox clergy could alone afford them protection and support in the dark ages of Turkish rule. But there are already indications of an approaching change. As in Bulgaria a religious movement preceded the political movement, so also among the Wallachs it is not improbable that the first step towards the assertion of a national existence may be the demand for a national Church. Coming events cast their shadows before, and a slight incident which occurred a few weeks ago in a Wallach village of the Pindus may perhaps be pregnant with big results. One Sunday morning the villagers collected round the church and informed their priest that he had to read the Mass in Wallach. The priest, though himself a Wallach, demurred to their request, probably because, like many of his flock, he scarcely understood the sense of the Greek words which he recited. The peasants, however, insisted, and replied that they were one and all determined not to attend again at church until the service was performed in their own tongue. On that Sunday the priest celebrated the divine service in an empty church. In the course of the week he thought better of it, and on the following Sunday Mass was read in Wallach, probably for the first time in history. Since then the priest has been suspended, the village threatened with excommunication, and the church closed by the ecclesiastical authorities. If, as the Greek clergy assert, this incident was merely the work of Roumanian propagandists, the movement may be nipped in the bud; but if, as I believe, it is the first open expression of a general feeling among the Wallachs, the spiritual and temporal weapons of the Orthodox Church will be as powerless to check its development as they proved themselves to hinder the growth of the national Church of Bulgaria. It has always been the error, not to say the crime, of the Patriarchate, that it has invariably failed to allow sufficient latitude for national expansion within the bosom of the Orthodox Church. It was not till after many years of struggle and the threat of a disastrous schism that the Patriarchate recognised the claims of free Greece to a national autokephalous Church. The same obstinacy has provoked the Bulgarian schism, and has, by estranging the Slavs of Turkey, endangered the very existence of Hellenism. Such another criminal folly may lose to Greece the wavering allegiance of the Wallachs, for their sympathy at the present moment is of but a negative sort. Their nation does not probably count altogether more than half a million of souls, and it is only through annexation to Greece that they have hitherto looked for release from Turkish rule. But military conscription, heavy taxation, possible restrictions on their nomadic habits, are so many circumstances which tend to temper their enthusiasm for Greek annexation. Were the headstrong pride of the Greek clergy to involve them in a struggle with their spiritual leaders, the chief tie which binds them to Greece would snap. Roumanian propagandism is already at work amongst them; and though in some places its emissaries have been coldly received, in many others their words have found ready listeners, for they have been addressed to men among whom the pride of race is as strong, if not stronger than among any other of the rising nationalities of Turkey, and the experience of the last few years is at hand to show how quickly and irresistibly the spark concealed for so many centuries under the smouldering ashes of past traditions can leap once more into an all-consuming flame.

This propaganda has generally taken the form of an educational propaganda; and though its efforts have been mainly directed towards Macedonia, as it was there that it seemed the most urgent to counteract Hellenic influence, and the subject might therefore have been properly alluded to in connection with that province, I refrained from doing so until I had dealt with the Wallach question in general.

This movement is not altogether a new one, as seems often to be assumed. It originated sixteen years ago in the foundation at Bucarest of a Macedo-Roumanian Syllogus for the encouragement of national education among the Romounoi south of the Danube. Substantial pecuniary assistance has always been forthcoming in aid of the undertaking to which the political events of the last few years have given a fresh impetus. In 1880, no less than £4000 was expended by the Syllogus; and I understand that this year that sum will probably be exceeded. Up to the present time fifteen schools, of which three are for girls, have been established under the auspices of the Syllogus, who has sent out all the teachers from Roumania (sixteen male and four female). The boys' schools are attended altogether by 1200 pupils, the girls' schools by 250; and notwithstanding the opposition of the Greek clergy, who denounce them in the blackest terms to the Turkish authorities, they are gradually moulding by their teaching and influence the minds of the Wallach youth.

When, however, it is asked what good will result from this propaganda for the future of the Wallachs, it is difficult to find any satisfactory answer. They are united by so many ties to the Hellenes, that their future amalgamation with Greece has hitherto been looked upon as an almost foregone conclusion. Yet the primary object of this propaganda is to render that amalgamation more difficult. They are not, however, numerous enough to form an independent State. Union with their trans-Danubian cousins appears to be a geographical impossibility. The propaganda must necessarily work equally against their absorption either by Albania or Bulgaria. The movement is therefore only a negative movement, and as such appears doomed to exercise no decisive influence on the destinies of the Wallachs.

# CHAPTER X.

## YANINA.

AUTUMN, with its constant alternations of fierce storms and bright sunshine, is the season which most harmonises with the wild nature of Epirus. The swollen streams rush headlong down its narrow valleys, or leap over its cliffs in foaming cascades; the wind sweeps freely over its bleak precipices; the forked lightning plays among its lofty peaks; the thunder rolls in resounding peals from rock to rock; while now and again the sun bursts forth, shedding rainbows on the retreating clouds, and lighting up with its transient glory the grand outlines of the desolate landscape.

After the first sharp descent from Metzovo, the track to Yanina—for it would be idle to call it a road—lies for some fifteen miles along the bed of the Arta. Yet in olden days there must have been a road, and a fine and frequented road too, to judge from the number of *hans* or way-side inns, and bridges, of which traces are still to be seen. But the *hans* are nowadays empty and ruinous, and the bridges are only marked by broken piles, or by one solitary span spared by the devious torrent. Needless to say that in winter, communication is constantly interrupted by snows and storms, and Epirus and Thessaly are temporarily cut off from one another. On both sides, the mountains rise grandly from the broad bed of the Arta Metzovitico, which is presently joined by the sister waters of the Arta Zagoritico, descending from the heights of the Zagori district. Villages nestle in the shelter of their flanks, looking at this distance prosperous and peaceful enough amid their oases of green trees and terraced fields. Yet there is scarcely another district in Turkey where lawlessness and brigandage have wrought such havoc as in Zagori. For the last three years it has been the happy hunting-ground of two formidable Greek bands, led by two brothers, Davelli by name, who have acquired for themselves in Epirus no less infamous a reputation than the mighty Kapitanos Kathrakia in Macedonia. The highlands of the Zagori district contain forty-three villages, all Christian; within three years more than half of them, and those the wealthiest, have been burnt, pillaged, and desolated by these ruffians. Nor are they content to spoil them once and for all; time after time do they return to the charge, carrying off the wealthier inhabitants for ransom, outraging the women, plundering and destroying in the mere wantonness of lust. In the presence of these atrocities what does the Government do? Nothing, worse than nothing. Now and then an expedition is organised, and a troop of Circassians sent off to the hills to find the brigands; but whether the latter receive information from the terror-stricken peasants, who fear their revenge, or from corrupt officials, who take their bribes, they always succeed in eluding pursuit, and the Circassians, after living for a few days at the villagers' expense, and completing their ruin, return to their quarters until a fresh opportunity is afforded for another such fruitless errand. The police alone are sometimes more successful; for in this province, where they are, almost without exception, recruited from among the Albanians, they are, as a body, honest and energetic. Only nine months ago a detachment of *zaptiehs* was despatched against the notorious band of Leonidas; and after six days of ceaseless marching and countermarching, they tracked the brigands to their lair, and the colonel of the *zaptiehs*, who was himself in command of the expedition, slew the robber chief with his own hand. But the *coup* has never been repeated. The *zaptiehs* are neither sufficiently numerous nor organised to cope with so gigantic and widespread an evil, and the military authorities, who perhaps alone have the power, are criminally supine. Mehemet Zekki Pasha, the general in command of the troops in Epirus—a fanatical and haughty Circassian, who owes his high and rapid promotion to Palace influences—is reported to have said that, "so long as Greeks only killed Greeks, the harm was not very great." But of late the brigands have not restricted themselves to the innocent amusement of torturing unfortunate peasants—they have even ventured to attack some of the leading Mussulmans; and no little commotion was caused some ten days before my visit to Yanina by the murder of Muslim Agha, one of the prominent promoters of the Albanian agitation. It remains to be seen whether these outrages committed on his own co-religionists will rouse his Excellency from his supineness, or whether indeed, as many of the Mussulmans assert, his indifference is but a cloak conveniently assumed to disguise his incapacity or worse. But, whatever the reason, the unhappy peasants of the Zagori are abandoned to the tender mercies of the brigands, and their once thriving villages are being fast converted into desolate ruins. More than 600 of them have already fled from their homes, preferring misery and exile to the horrors of their present existence.

For three hours our route lay along the border of the Zagori district; and the groups of mournful wanderers journeying towards Yanina, with the scanty wrecks of their household goods and chattels, bore eloquent testimony to the stories which our escort had to tell of the Davellis' savage bands. After crossing and recrossing some twenty times the tortuous bed of the Arta, our track suddenly left the valley to climb the steep ridge of Mount Drysco, which alone separates the main chain of the Pindus from the plain of Yanina. An hour up the zigzag path, carried away in many places by landslips, and we stand on the sharp crest of the hill. At our feet lies the lake, overcast at present with the reflected gloom of heavy storm-clouds; the precipices of Mount Metzikeli rise sheer out of the bosom of the glassy waters, but its lofty peaks are lost in darkness above us; over on the other side, the grim battlements and yawning vaults of the ruinous fortress, which still attest the barbaric pomp and power of Ali Pasha, and the town, with its many domes and minarets, and streets made bright with white houses and green trees, are still lighted up with a furtive ray of sunshine; but to the north, beyond the island where the octogenarian tyrant met his fate, the head of the lake and the distant mountains of Albania are veiled in blackness, lightning quivers in the clouds, and the thunder rolls incessantly, like the din of distant battle. Long ere we reach the foot of the hills, the strong blast of the storm-wind has swept over the lake, lashing its waters into white fury; and the tempest of rain which beats upon us as we plod for three hours across the marshy swamps which border the southern end of the lake, thoroughly confirms at least the last portion of the Turkish saying, that at Yanina there are only three things worth seeing—the lake, the tomb of Ali Pasha, and the rain. But Yanina rains are short-lived. Ere we enter the walls of the town the sun once more glistens merrily on its bright stone houses and tiled roofs; the storks, perched on their high nests, are shaking their rumpled feathers; big rain-drops sparkle on the trees which overhang the walls of many a garden and mosque; and the neat, well-paved streets are crowded with the picturesque bustle of everyday life.

If sentimental considerations were alone to decide the fate of Yanina, there would be little room for questioning the propriety of the decisions of the Berlin Conference. There are few districts, even in Greece itself, where the continuity of Hellenic culture is so indisputably established as in the town of Yanina and its surrounding region. Here, indeed, is to be traced the very origin of that splendid name which inspired of old the valour of the soldier, the eloquence of the orator, the divine breath of the poet, and the creative power of the architect and sculptor, and which, at the present hour, if it serves too often to colour the paltry ambitions of restless agitators, still embodies the best aspirations of the most progressive section among the Christian populations of the Levant. Since the days when, according to tradition, Hellen, the son of Deucalion, made the sanctuary of Dodona sacred to Zeus, and bequeathed the name of Hellopia or Hellas to the small highland kingdom over which he ruled beneath the western slopes of the Pindus, Hellenism has pointed the two extremes of human fortune; but through all the vicissitudes of history, it has never ceased to live as the civilising and vivifying element in its native region. Yanina itself may fairly be looked upon as the not unworthy daughter of Dodona. Though recent discoveries have removed the site of the ancient sanctuary of Jupiter and Dione from the shores of the lake of Yanina to the vale of Merovishta, some eight miles to the south-west of the modern town, there is little doubt that, when religious and political revolutions wrought the destruction of Dodona, it was the inhabitants of the sacred city who, driven from their homes, founded the town of St John ('Ιωάνινα) on the fertile banks of the lake which now bears its name; and during the five centuries of Turkish rule, when Athens and Corinth were mere Turkish villages, the descendants of the Hellenes have never ceased to resort to the schools and colleges of Yanina, even as their ancestors had flocked to the national shrine of Dodona.

The influence of Hellenism in Yanina must not be measured by simple figures. The whole Sandjak of Yanina contains, indeed, an overwhelming majority of Christians,—the districts of Zagori, Kourenta, Malakash, Tchioumerka, and Tcharkovishta showing a population of 130,000 Christians (Greeks and Wallachs in almost equal numbers) scattered over upwards of 200 villages, where the Mussulman element is unrepresented save by the landowners and their immediate households. The town of Yanina, with a population of 20,000 souls, counts only 11,500 Greek Christians as against 5000 Mussulmans and 3500 Jews. But the

Greek language has retained undisturbed its ancient supremacy among all sections of the population. The Moslem coming out of mosque, and the Jew returning from his synagogue, speak the same tongue that is heard from the pulpits of the Christian churches. The number and excellence of the Greek schools cannot suffice to account for this phenomenon. The influence of a highly developed educational system, which is represented by nine popular schools, besides a theological seminary and two high schools in the town, and 198 village schools throughout the district, must necessarily permeate the whole population. But it has operated not to introduce, but only to preserve, the Greek tongue. Christians and non-Christians equally speak Greek, because they are equally of Greek descent. Among the Mussulmans there is but a small strain of Asiatic blood. The history of the Mussulman annexation of Epirus is curiously illustrative of the process by which the Byzantine rulers worked out their own ruin. During the wars and revolutions which convulsed the rotten fabric of the Eastern empire, a branch of the imperial house of Comnenus had succeeded in appropriating Epirus, and reigned over it with a few brief interruptions as independent despots for the space of over one hundred years. But Andronicus III., after a successful campaign against the Servians, turned his victorious arms against the despotate, and reduced the province to subjection. His success was, however, mainly due to the assistance of a body of 2000 Turkish mercenaries, and, in return for their services, he was constrained to grant fiefs to their leaders of the richest lands of Epirus. Henceforth, whenever a Servian or Albanian invasion threatened the country with ruin, whenever rival chiefs were contending for sovereignty, it was the action of these Turkish chieftains, often backed by the armies of the Sultan, which struck the balance; and bands of Ottoman mercenaries were always to be found, fighting now with one party and now with another, but strengthening always the prestige and power of Islam. At last in 1431 the despot Charles, son of the Kephaloniote usurper Esau, unable to reduce the revolted city of Yanina, entered into direct negotiations with the Sultan Amurath, and agreed to surrender to him the whole province of Epirus on condition that he should be put in possession of Yanina, and allowed to hold his court there as a vassal of the Ottoman throne. Six Turkish chieftains were despatched to take over the new province, and Epirus was definitely incorporated with the Ottoman empire. At first it was treated by the Sultans with conspicuous leniency; its ancient privileges and immunities, solemnly confirmed at the time of the annexation, were respected; and the practice of the Christian faith was absolved from the burdens with which it was fettered in other parts of the empire. The Epirote landowners continued to enjoy possession of the fiefs which had been granted to them by former despots and Byzantine emperors; and though the military service to which they were liable under the conditions of their original acts of allegiance was rigorously exacted by their new suzerain, they were allowed to form a separate corps, and fight side by side with the followers of the Crescent, under their ancient standard of St George. But the power which they enjoyed at home, and the valour which they displayed upon the battle-field, at last roused the jealousy of their Mussulman sovereigns, and the Spahis were compelled, under threat of confiscation of their *spahiliks* or military fiefs, to embrace the creed of Islam. The mass of the Epirote landowners acquiesced in the new conditions, and surrendered their faith rather than forfeit their lands. The Mussulman beys of Epirus are the descendants of those renegades, and they have retained to the present day the manners, type, and language of their ancestors. But, although the influence of their Hellenic origin and surroundings may still be traced in the prevalence among them of a more progressive spirit than is generally to be found among Turkish Moslems, it would be idle to pretend that that influence is sufficiently strong to reconcile them to the prospect of Greek annexation. Though they are as a body the first to admit the inherent vices of Turkish administration, they enjoy far too large a share of local power under the present system to look forward without apprehension to the effects of Greek centralisation on their own position. No doubt, besides that fair share of local influence which the apathy rather than the intelligence of the Ottoman Government allows in certain provinces of the empire, what the Mussulman bey in many cases dreads to lose is the licence and indulgence too often extended to his abuses. But among the more upright and intelligent—*i.e.*, among those who most ardently wish for a change from the existing state of anarchy and incompetence—the repugnance to Greek rule is general and genuine, and deserves consideration, because it is founded, not on religious hostility, but on the clear appreciation of the defects of a governmental system which has already worked so ill within the present limits of the Hellenic kingdom—viz., excessive centralisation—which hands the provinces over, bound hand and foot, to the tender mercies of the representatives of the central Government, and the constant struggles of factious parties, who look upon the provinces merely as the prizes of war, or as bribes with which to secure peace. This feeling exists even, in a lesser degree, among those whose eyes have been turned from childhood towards Greece. I will not allude again to the Wallachs, but even among the most thoroughly Hellenised Epirotes, among men who have smuggled their children out of the country to fight for Greece, with all her faults and imperfections, there is a growing conviction that, if Greece is to carry out worthily in these parts the civilising mission which they ascribe to Hellenism, she must abandon the narrow traditions of a despotic bureaucracy. She must, moreover, guarantee to each locality at least that measure of provincial and municipal self-government which the Turkish Government has been forced to tolerate, and give a distinct earnest that Greek rule means something better than the substitution of the Nomarch for the Pasha, and the caprice of Athens for the tyranny of Stamboul.

Among the various sections of the population of Yanina, there are none who look forward to Greek annexation with more dread and bitterness than the Jews. Descendants, they claim to be, of a colony of Israelites who emigrated to Epirus even before the Christian era, some say as early as the second captivity. Equally removed by religious differences from Christian and from Moslem, they have had time during two thousand years to compare the persecuting fanaticism of the former with the tolerance, however contemptuous, of the latter. The traditions of their sufferings under the rule of native despots and Byzantine emperors have been constantly revived by fierce outbursts of Christian fanaticism. It is only seven years ago that a Jew was torn to pieces in the streets of Yanina by the Greek mob; and a system of social ostracism was inaugurated under the auspices of the Greek clergy against the whole Jewish community, until the Turkish authorities were obliged to interfere and coerce the Christians into at least the outward observance of their own evangel. If such incidents are not calculated to enlist the sympathies of the Jews on behalf of Hellenic aspirations, there are also other causes which operate to make them view with repugnance any modification of the present state of things. There are many peculiarities connected with Turkish rule which appeal to the money-grubbing instincts of the Hebrew community. The impecuniousness of the Government, which furnishes opportunities for advancing loans at the most usurious rates of interest, the buying and selling of the tithes, the constant fluctuations of a currency at the mercy of vizierial caprice, the very distress to which the country is reduced by the haphazard misrule of its governors, are so many circumstances which the Jew full well knows how to turn to the best account. Whether he consult his conscience or his pocket, he has no cause to wish for a change, but rather to bless the present system, which allows him to serve both God in peace and Mammon with profit.

It is a relief to turn away from the bitterness of party spirit, and the jealousy and suspicion which convert the various communities of Yanina into so many hostile camps, to the calm and eternal beauty of its natural surroundings. Yanina no longer covers the same area as when Ali Pasha, a king in all but the name, adorned it with all the pomp of his barbaric court, and swore by his long white beard that his capital should equal Stamboul in magnificence as it already rivalled it in power. The massive castle which he built out into the lake is but a heap of shapeless ruins; no trace is to be seen of the splendid palace which astonished even Hobhouse and Byron; the gunboats which he brought overland from Prevesa at the cost of untold labour and expense no longer darken the peaceful surface of the lake; the wall-like slopes of Mount Mytsekeli no longer re-echo the sounds of nocturnal revelry, or the wild shrieks of victims and of dupes committed to the safe keeping of the deep waters which slumber at its feet. But the town which has risen out of the ashes of Ali's capital still offers much to charm the eye and interest the mind; and, if the traveller wearies of the picturesque forms which animate its streets and quaint bazaars, he can pass out from its crowded thoroughfares to the gardens and groves which cluster on the neighbouring hillside, and watch the varied forms of hill and dale, the opposite precipices of Zagori, the gentle slopes of Drysco, the distant peaks of Pindus grow crimson under the embrace of the setting sun, and pale to ashen whiteness as the fickle luminary hurries away on his westward course.

# CHAPTER XI.

## "BY SULI'S ROCK."

STARTING is always a laborious effort in the unpunctual East. There seems no end to the small delays. First, a muleteer does not put in his appearance, then a servant has forgotten some article of alimentary necessity, and finally, perhaps there is a difficulty with the Pasha about the escort. For, alas! there are now few provinces in Turkey where one can dispense with the latter *impedimenta;* and in Epirus of all other provinces brigandage is rampant. The sun was already high in the heavens when our cavalcade defiled through the streets of Yanina. The good townsfolk were just streaming out of church (for it was Sunday morning), in time to gape at our procession, which consisted of five *zaptiehs* or mounted police as escort, myself and another English traveller, two servants, a Circassian and an Albanian, and a tail of pack-animals and muleteers. We sallied out by the Prevesa and Arta road, and as we rose over the first range of hills which bounds the Yanina plain to the south, we had a lovely view of the bright little capital of Epirus, its white houses gleaming amid the green foliage of many gardens, the dark walls of Ali Pasha's ruined castle projecting into the blue lake, the smooth waters studded with fleet sails, the little island where the revolted satrap was shot down in a Greek convent by the Turkish soldiery, the wall-like precipices of Mount Mytsekeli, and far beyond, the delicately sculptured crest of the Pindus already capped with winter snows. Our track soon diverged westward from the main road, and after climbing one or two steep ridges, we descended into the small secluded valley of Tcharakovista.

The site of Dodona was, until lately, a much-vexed question; and when Byron asked—

> "Oh where, Dodona, is thine aged grove,
> Prophetic fount and oracle divine?
> What valley echoed the response of Jove?
> What trace remaineth of the Thunderer's shrine?"

—there was none to answer him. But the excavations begun in the valley of Tcharakovista by M. Mineyko, a Polish engineer in the Ottoman service, and continued by M. Karapanos—the statues, inscriptions, and other relics unearthed by those gentlemen's exertions—have, it is generally admitted, finally settled the question, and the traveller who enters this peaceful little vale knows that he is treading the same holy ground to which the pilgrims of Hellas wandered with pious steps three thousand years ago. Dodona was the earliest, as it was also the most venerable, of Hellenic sanctuaries. Long before the development of Greek polytheism, a national shrine existed there, embodying the primitive conception of the Divinity which created and maintained the universe. There it was that, at a later period, the supreme force which formed and ruled the world was personified in Zeus, who received as emblems the thunderbolts that shattered the crests of Mount Tomaros, the eagle that nested in its ravines, the oak that grew on its flanks; while the water which gushed out of its rocks to permeate and fertilise the soil, formed another of his attributes, and gave him the surname of Naïos. Then the productive earth was identified with Dione, the primitive spouse of Zeus; and Love, the fructifying element in nature, was represented by Aphrodite, the daughter of Zeus and Dione, whose emblem was the dove, the sacred bird of Dodona; while to the genius of Destruction and of Death were assigned the precipices of the southern slopes of the Tomaros, where the Acheron led through inaccessible ravines to the infernal regions. Little wonder that the shrine which thus embodied their faith in the higher powers that rule all things should have been the resort of Greeks from all parts of Hellas whenever issues of great moment were at hand. To the oracle of Dodona Inachus goes for advice when Io has related to him her dreams. The priestess of Dodona foretells to Io her relations with the father of the gods, and to Hercules the end of his labours. To Zeus Dodonaios Achilles appeals to protect his beloved Patroclus: to him Ulysses journeys to learn how he is to return to Ithaca. Orestes, the son of Agamemnon, consults his oracle; and Pyrrhus, the son of Achilles, weds Larissa, the niece of Hercules, before his altars. Æneas and his Trojans, who have experienced in the fate of war the strong hand of the Greek gods, travel to Dodona to inquire where they are to found their new colony. Verily the shrine which thus sanctified the laws and actions of kings—which predicted the issues of wars, and gave counsel in all things, private and public —which informed the ideas and feelings of a primitive nation, and often maintained union where there was no bond of political unity,— may still be approached in a spirit of reverence, even by those who would say with the French poet—

> "Je suis venu trop tard dans un monde trop vieux,
> Où d'un siècle sans foi naît un siècle sans crainte."

But the days are now far away when from all towns of Greece offerings flowed in to the temple of Zeus amid the splendid pomp of a solemn ritual, and Demosthenes, addressing the representatives of Dodona, welcomed them as the ambassadors of a city which he esteemed of all cities the most blessed and the most beloved of gods, for that Zeus Dodonaios and

Dione and the Pythian Apollo spake through the mouth of its oracles, and showered their favours upon the town. There is little left, save confused heaps of stone, of these once world-famous temples. A massive wall of early Hellenic structure, however, still marks the limits of the town which grew up beside the sanctuary, and bushes of stunted ilex have taken the place of the sacred groves where pilgrims were wont to wander in a dim religious twilight. One monument alone has escaped the ravages of time and the fiercer hand of man. The theatre, where the Naïan games were celebrated in honour of Zeus and Dione, still remains in comparative preservation, one of the best specimens among similar relics of ancient Greece. It is built against a natural hollow in the hillside, just beyond the walls of the town, and its fifty rows of stone seats are supported on masses of carefully cut, uncemented stone-work. Hence the spectators could survey the temples and groves of the sanctuary, the lofty peaks of Mount Tomaros (now called Mount Olytzika), which still bears on its flanks black patches of oak forests, and through the southern mouth of the valley the distant precipices which overhang the Acheron.

Time slips away too fast amid the memories of such a past; and when the captain of our escort warned us that, if we tarried longer, we should not reach our destination before nightfall, it was with a sigh of impatient regret that I mounted my horse and turned my back on the spot where Dodona was. We had taken letters of recommendation from a kind old Albanian Bey of Yanina, Djemal-ed-Din Agha, who owned several large *chiftliks* in this part of the country, and had hospitably insisted that we should make two of them, at least, our resting-places for the first and second nights of the journey. We had to double back a little bit to the north, and follow the valley of Tcharakovista to its northern extremity in order to round the lofty heights of Mount Olytzika. The gorgeous tints of sunset faded into twilight while we crossed and recrossed the bed of a tortuous stream, which seemed to do duty for a highroad, and ere we reached the hill on which our *chiftlik* was perched it was already almost dark. Its name was *Eleutherokhoria*, or Free Village, but its freedom, like that of many others, had ceased in the days of Ali Pasha, and it was now the property of the Albanian Bey. The rising moon helped us to struggle, without too much stumbling, over the last piece of the road, and when we arrived on the small plateau at the summit of the hill, a bright fire flaming in the courtyard outside the house, and the grateful spectacle of a sheep impaled on a big wooden spit, announced that our arrival had been duly heralded, and becoming preparations made for the reception of the tired and hungry travellers. The headman of the farm came out to welcome us, and we were soon comfortably installed with soft rugs, a blazing fire, and good substantial cheer.

Dawn was just breaking into daylight as we sallied forth the next morning from the *chiftlik.* It was one of those rare autumnal days when Nature, arresting for a moment her process of annual decay, flings once more over the world her departing loveliness in exuberant defiance of the frosts and storms

> "Which dreary winter leads
> Out of his Scythian cave."

From the summit of the knoll we could now survey the landscape which on our arrival was already wrapped in the gloom of fading twilight. Far away to the east the snow-capped peaks of the Pindus floated above a silver mist in a sea of rosy light. Straight in front of us, to the south, the frowning mass of Mount Olytzika rose gaunt and bleak out of a maze of blue hills and vales, wearing on its shoulders a purple mantle of dark forests, and on its brow a crown of burnished gold. To the west the jagged outline of the mountains of Suli, flushed with the warm embrace of the rising luminary, stood out in brilliant relief against the transparent heavens. But our route lay down in the valley, and we soon found ourselves shut out from the prospect of the distant mountains by two ranges of low wooded hills, the outworks of Olytzika and Suli, which ran along both sides of the valley. We were not, however, inclined to cavil at the change. The sun, as it rose higher in the sky, beat down upon us with scorching power, and the spreading plane-trees, which stretched their canopy of yellow foliage above our path beside a brawling torrent, afforded us grateful compensation for the breezes we had left behind us on the heights, while there was food enough for the eyes in the gorgeous tints which clothed the wooded slopes around us. There was scarcely a village or a human being to be seen during the long day's ride. Only a ruined water-mill or a deserted *han* showed here and there that the hand of man had once been busy in these regions. Once, through a break in the hills, we descried on the crest of a rocky bluff what looked like a Greek convent. I inquired of one of our *zaptiehs* what it was, and received the same everlasting answer, "*Harab dir.*" (It is a ruin). "What ruin?" "*Kim bilir!*" (Who knows!) We halted for our noonday meal under the shade of a giant oak-tree, "The tree," as it is called in the neighbourhood—for in this region, which was once covered with forests of oak, there is nothing left but brushwood, and the maple and poplar trees, which grow out of the exuberance of the soil wherever there is water to bathe their roots. Some supersti-tious reverence seemed to have saved this one relic of the ancient forests; but lightning, which knows not such fears, had shattered half its branches and seamed its colossal trunk, thirty-five feet in circumference. A few miles beyond it the valley gradually expanded. To our right towered above us the sheer cliffs of Suli, and on a projecting spur the white walls of the large fortress-looking *chiftlik* where we were to find our night's quarters. The barking of the deep-mouthed Molossian dogs, who, from afar, heralded our approach, broke gratefully on the desolate silence which had surrounded us throughout the day; and as we climbed the last slopes of the hill, the village of Romano itself came into view, among fields of maize just ripe for the reaper, and long tracts of rich brown land furrowed by the plough. Here everything was once more informed with life. From the steep mountain alps above us, herds of cattle, returning to their night's quarters, were making music with their bells. Large flocks of sheep and goats were loitering on their way home, notwithstanding the guttural solicitations of Albanian shepherd boys. Even the dingy hovels of the peasantry were made bright with the red and blue dresses of the women and the harsh clatter of many voices. At the gate of the outer wall of the *chiftlik* our host stood ready to receive us. It was, indeed, more of a fortress than a farm-house. A long wall, furnished with small round towers, and pierced with loopholes, surrounded the dwelling. Towards the mountain, whence alone it could be assailed, the house had nothing but blank walls, loopholed at every storey. All its windows, iron-barred, looked out on the valley, which the precipitous cliffs seemed absolutely to overhang. Thick walls of grey limestone, big rafters of solid timber, staircases of stone, were so many indications of the warlike forethought which had presided at its construction. It had belonged of old to a Suliote chieftain, and had been burnt out in the wars of extermination which Ali Pasha fought against the fierce mountaineers of Suli; but he had himself rebuilt it, and after his fall it had passed, with the rest of his enormous possessions, to the Sultans, who had sold it to Djemal-ed-Din Agha's father. The whole of the population had disappeared, either killed or in flight; but the new owners imported fresh labourers, Albanians from Konitza and Greeks from the plain of Yanina, and succeeded, in the course of time, in restoring some measure of prosperity to this desolated district. Fresh villages had sprung up out of the ruins of the old ones; and our host, a younger son of Djemal-ed-Din Agha, who had been for fifteen years in charge of the property, told us with no small pride, which was not unwarranted, that he had now three hundred families on these estates. The young squire certainly did not look like the son of a wealthy landowner who could count his revenues by thousands. His somewhat ragged clothes, in spite of their Frankish build, and more still, his hard rough hands and bronzed complexion, told us plainly enough what he himself repeated in words—"I am a labourer amongst labourers;" but there was still about him that unassuming dignity and quiet grace which are the distinctive features of a Mussulman gentleman. And as its master, so was the house. It was the home not of luxury, but of work. The large compact building was almost entirely devoted to the requirements of the farm. There were scarcely any outhouses, with the exception of a large roomy stable. Everything was stowed away in the central block; granaries, store-rooms, a large hall for the retainers, took up three-fourths of the available space, leaving only a small wing for the squire and his family, and the *selamlik*, or guests' room. The latter was neither spacious nor luxurious, its sole furniture consisting of a tattered carpet, and a long divan which ran round three sides of it, and was covered with a faded chintz. But nature made up for these deficiencies by the glorious prospect which the windows commanded over valley and mountain. Opposite to us was the massive ridge of Mount Olytzika, more naked and more precipitous even on this side than towards Dodona, while on its flanks the outline of the rugged range of Suli was cut out in sharply projected shadows. Out of a deep depression in the chain, the silver streak of the Acheron could be seen rushing headlong into the deep valley beneath us, whence the blue smoke of many a hamlet rose in wreathed columns out of the evening mist. The sound of song and laughter from the courtyard in front of the house announced the preparations for our dinner. A huge fire of dried branches was blazing against the wall; and by the side, just out of reach of flame and smoke, the customary sheep was being roasted on a wooden spit, flanked by two turkeys, impaled according to the same fashion. Three stalwart cooks superintended the operations, while the rest of the retainers crouched about the court, watching with interested curiosity the preparations for the unwonted feast. Now and then one or other of them would spring forward to relieve one of the officiating trio, who were themselves in no little danger of being roasted with their charge; but the spit never ceased to revolve, keeping time with the rhythmical chant of the assembly. The glare of the fire played fitfully on their dark sharply-cut faces and flowing hair; it gleamed on the bright weapons which stuck out of their belts; it glowed on their grey cloaks and *fustanellas*, and cast grotesque shadows on the walls, to the great delight of a deaf and dumb "simple," who vainly addressed them with eloquent gesticulations. He was a cousin of the house, and its spoiled child:

his wayward antics seemed to provoke rather reverence than laughter; for is it not written, "Blessed are the poor of spirit"? and did not Mohammed himself say, "Such as these ye shall treat like children; for verily they are God's own children"? But the most interesting part of the dinner — viz., the preparing of it—was terminated; and after the tamer, if not less necessary, process of consuming our substantial meal was over, mattresses and blankets were speedily produced, and we were left to our dreams, even before the fumes of the banquet had vanished from the walls of our apartment.

The next morning we were again off betimes; for we had a long and steep climb before us. We struck straight up the hills behind Romano, and the goat-track which we followed soon became so precipitous that we were fain to dismount, and leave our horses to struggle as best they could over the slippery rocks of shelving limestone; nor were we able to ride any more that day. If the ascent was sharp, it was at least rapid; and after clambering for three hours over rugged cliffs, and through thick underwood, we found ourselves on the sharp knife-like crest of Suli, nearly 4000 feet above the level of the sea. The western face of the chain looked even more precipitous than the slopes we had just scaled: there was scarcely the trace of a track to be seen—only the ruined homes of Kako-Suli, on a projecting ledge, some 2000 feet below us,—the long line of castellated rocks overhanging the gorges of the Cocytus and the Acheron, which were once the strongholds of the Suliotes, and the Turkish fort, which still mounts guard over their ruins,—marked our destination. Whether we had missed the proper track, I know not; but as we scrambled down the steep ridge, and our horses and pack-animals stumbled, and tumbled, and groaned over the sharp rocks and treacherous brambles which concealed their pitfalls, I was forcibly brought to the conclusion that when Virgil wrote "Facilis descensus Averni," the poet delivered himself of an exceedingly misleading assertion. There were, however, no bones broken; and after a couple of hours of anything but facile descent, we were safely landed on the small mountain plateau where the "ghost of freedom haunts" the silent ruins of ill-fated Suli. Dismantled, fire-scarred walls alone mark its once prosperous homesteads; but the terror of its name still lives in the popular legends. When I proposed to pitch our tents for the night amid its ruins, there was one cry of horror from guards and muleteers, "Did I not know that every night the spirits of its last defenders hover about the air, making the dark hours hideous with wild war-songs and lamentations; and that no mortal can hear the fierce shriek with which they vanish at the first break of dawn, without dying? Had not old Dimitri, the miller, been wandering about the hills with scattered senses ever since the night when the storm drove him to take refuge under Suli's accursed wall? *Olmaz, Effendim! Olmaz!* It cannot be!" There was no overcoming their stubborn fear; and so we made for the Turkish fort, where at least I could depend upon a hospitable reception. Alas! Mr Gladstone's policy had wrought its effects even in these secluded regions. At the gate of the castle, the captain in command of the 100 men who form its garrison met us with deep protestations of regret that, three days ago, he had received stringent orders from the military authorities not to allow any strangers within the precincts of his fortress. "We had a *bougourouldu* from the Governor-general of Yanina. He was ready to kiss the Vali Pasha's seal; but unfortunately he was an officer, and could only obey the orders of the military authorities. Orders were orders; and however profoundly pained to deny us ingress into his own house, *Olmaz, Effendim! Olmaz!* It cannot be!" However, at the foot of the castle-hill there were two hovels, which we had passed on our way up, and where, we were told, there lived the descendants of the only Suliote family which had been spared from the national destruction; and there we might hope to find food and shelter from the storm, which was already gathering about the Ridge of Lightning, as the highest ridge of Suli is called. But before retracing our steps, we had to perform our pilgrimage to the ruins of the fort where the last and most terrible scene of the Suliote tragedy was enacted.

The Suliotes were a hardy race of Christian Albanian highlanders, who, by their warlike prowesses, maintained during the last century undisputed supremacy in these mountain fastnesses, and often carried terror down into the plains, as far as the sea on one side and the gates of Yanina on the other. Ali Pasha, the famous tyrant of Epirus, after breaking the power of the feudal chieftains of Albania, turned his might against the handful of bold warrior-peasants, whose very independence was in his eyes a sufficient offence. But for fifteen years they defied his attacks, and many a time was the proud Pasha compelled to fall back defeated from their inaccessible heights. Where brute force had failed, diplomacy in its most usual oriental form—viz., bribery—prevailed. George Botzaris, the most able and powerful of the Suliote chieftains, was bought over, and a fresh campaign planned by the traitor himself, led slowly but surely to the final catastrophe. Though the Suliotes were never numerous (at the zenith of their power they could not muster more than 3000 fighting men), the small highland plateau of Kako-Suli was incapable of providing for their wants. Ali Pasha, instead of hurling his strength in vain assaults on their natural strongholds, simply blockaded them within the narrow limits of their barren mountain home. Cut off from the plains which they had been wont to harry at will, the Suliotes were reduced to the severest straits, and disaffection began to make ravages amongst their ranks. The clan which held the northern key to their position deserted bodily to the enemy; and gold, which had laid the foundations of their fall, was once more used successfully by Ali Pasha to gain ingress into Kako-Suli. A certain Pylios Gusi introduced the Pasha's troops under cover of the night into his house, and the next morning Kako-Suli was in their hands. Desperate, but undaunted, the Suliotes threw themselves back upon their last retrenchments. To prolong the resistance of the remainder, a large body of them cut their way through to Parga, and thence sought refuge in Greece, where they lived to pay off their vengeance tenfold on the Turks during the war of independence. The last band of defenders held the fort where we were now standing: in front of them every coign of vantage occupied by Ali's victorious troops, behind them the sheer precipices of the Acheron. Their position was hopeless; but the fierce eloquence of Samuel, the warrior-monk, inflamed their heroism, and when the worst came to the worst they preferred destruction to surrender. As the Pasha's soldiery scaled the last Suliote stronghold, Samuel and his devoted band retired slowly before them, fighting inch by inch; and when the last retrenchment had been stormed by sheer weight of numbers, a fearful explosion involved in the same ruin both victors and vanquished. Samuel had with his own hand laid the lighted match to the powder-magazine. Meanwhile the Suliote women had escaped on to the rocks which overhang the Acheron; and when the report of the explosion announced the final disaster, it is told that they raised their voices in a last chant of desperate triumph, and taking their children in their arms, flung themselves headlong over the precipice into the dark waters of the River of Death.

Seventy-seven years have passed since this awful tragedy was consummated, and a few blackened vaults overgrown with wild mountain shrubs alone mark these scenes of woe. *Solitu-*

*dinem faciunt, pacem appellant.* The shrill note of a bird of prey, or the harsh voice of the sentinel on the battlements of the Turkish fortress, alone disturbs the peace which nature sheds over the theatre of man's worst as well as noblest deeds. The calm of evening rested on the wide and varied landscape unrolled before our eyes. To the left of the southern chain of Suli, which rose immediately opposite to us, separated from the central ridge only by the deep chasm of the Acheron, lay the broad Ambracian Gulf backed by the distant peaks of Acarnania, and to its right the placid waters of the Adriatic, with Paxos and Antipaxos reposing on their bosom. But behind us heavy storm-clouds had enveloped the higher ridge of Suli, and the muttering of thunder bade us hie back to our night's quarters.

Nor did these turn out so comfortless as might have been expected from their external appearance; and besides, was not the fact of sleeping under the roof of the last of the Suliotes worth some slight measure of discomfort? How they had managed to escape the destruction which had overtaken the rest of their race was a question to which I failed to elicit a satisfactory reply. The three men who with their families tenanted the two small cabins, were the sons of two Suliote brothers who had borne a part in the tragical events of 1803, though what that part exactly was, they seemed either unable or unwilling to tell. All they knew, they said, was that their fathers rendered a great service to Ali Pasha; and that when the final catastrophe came, their lives, and those of a few women and children who had sought refuge in their house, were spared by the conqueror. They themselves had wedded their cousins, for Providence had kindly provided that each of the children of one of the original survivors could be mated with one of the other's children, and the numerous offspring with which the various couples had already been blessed, allowed one to hope that the race might yet sprout up to new and vigorous life. Though remnants of the Suliotes are still to be found scattered about the Peloponnesus and other parts of the Hellenic kingdom, they have lost by intermarriage with others that purity of breed which, it is said, enabled them to preserve down into modern times the perfect beauty of the Grecian type. And certainly both men and women in these humble hovels were worthy of their ancestors' high renown; the men, tall and powerfully built, with strongly marked faces, in which the deep lines and weather-beaten complexion, and shaggy beards and long dishevelled hair, rather enhanced than marred the bold sharply defined features. The women, more worn even than the men by toil and hardships, still bore witness to their high descent; and one young girl, who had only been recently married to the youngest of the three brothers, might well have stood for Milo's Venus, whether you looked at the graceful athletic figure, the delicate joints of hands and feet, the swelling bosom, the slope of the shoulders, and the small perfectly poised head, or at the classical features of the oval face. Nor were these charms unbecomingly set off by the peasant girl's simple dress: a small red fez, secured by a plait of hair wound round it, and fastened with a silver pin; a bodice of coarse blue cloth trimmed with red embroidered work, and a short petticoat of the same stuff, open down the front, but covered with a red apron, and leggings of quaint-coloured worsted down to the ankles. It was a curious group that gathered round the large blazing hearth, while the wind and storm were beating against the mountain-side. In the corner an old woman in her dotage, the grand-aunt of the family, who had witnessed with her own eyes the days of the "great trouble;" beside her five small children, confused bundles of picturesque rags, with bright eager faces and curly heads; and crouching opposite to the fire, while their wives stood spinning behind them, the three fierce-looking highlanders, whose eyes gleamed with the double light of the reflected flames and their own enthusiasm, as they recalled for the strangers' benefit the memories of bygone times when their fathers were the lords of the soil which they now tilled as hired bondsmen: and when the wind howled more wildly, or the thunder crashed more loudly, they would stop for a moment to cross themselves devoutly, and mutter a prayer for the souls of the Suliotes "who were about." Presently, when the storm lulled, all the grown-up members of the family retired to the other cabin, where our guards and muleteers were warding off the "ghosts" by uproarious song, and we were left to share the room with the children and the old crone, who never ceased all through the night to moan whenever a fresh gust of wind whistled through the rafters of the roof.

# CHAPTER XII.

## "AND PARGA'S SHORE."

In the morning the sky was once more bright and clear, and the big rain-drops on the trees and the brawling of the swollen torrents were the only signs left of the fury of the storm. A steep and slippery zigzag down the wooded slopes of the hill soon brought us to the dark ravine of the Acheron. The stream has carved itself a deep bed through the limestone rock which, gleaming through its pellucid waters, imparts to them a peculiar whiteness. On either side lofty cliffs tower over the water-course, — on the one side the precipice down which the Suliote women took their heroic leap into eternity, on the other the bulwarks of the southern range of Suli, which stretches away almost as far as Prevesa. Dark gnarled oaks growing out of every fissure in the rock, project their spreading branches above the stream, and cast over it a gloom which even the rays of a vertical sun can scarcely dispel. A little further on another torrent descends from the northern hills of Suli, and the two rivers of Hell, mingling their accursed waters, disappear in a wild gorge whence we could hear the roaring of a distant waterfall. Our track lay over the wooded ridge above it, and for more than an hour we toiled painfully through a tangled forest of oaks, and ilex, and arbutus, and wild mulberry, twined and intertwined with countless wreaths of luxuriant creepers, and bright with the rich clusters of their berries :—

> ". . . Arbuteos fœtus, montanaque fraga
> Cornaque et in duris hærentia mora rubetis
> Et quæ deciderant patula Jovis arbore glandes."

On emerging from this almost virgin forest we found ourselves on the brow of a cliff overhanging the plain. For beneath us the Acheron, issuing triumphantly from the precipices of Suli, flowed with majestic dignity across the lowlands towards the distant haven of Glykys Limen, where its waters sweeten the briny sea. Down the face of this cliff we now had to descend. The prospect was not inviting. It was to all appearances a sheer wall of rock, with here and there a patch of brushwood or a mass of loose rubble clinging to its interstices. But our guide said it was the Sultan's road, and people had travelled over it for generations—though, as he added in the same breath that he never recollected pack-horses having crossed the mountains of Suli, the latter statement somewhat detracted from the strength of his argument. We had, however, no option in the matter. So the *zaptiehs'* clever ponies showed the way, jumping like goats from ledge to ledge; then our own horses were driven down; and behind them, at a safe distance to prevent collisions, the pack-animals slipped and slid and stumbled, a *zaptieh* or a muleteer clinging on desperately to their tails and thus acting as a kind of break, —and finally all the beasts were safely landed on a little plateau, beyond which the descent became less precipitous. Having watched the exciting operation from the top, we proceeded to rejoin our caravan; and though we too had often to transform ourselves temporarily from bipeds into quadrupeds, the *mauvais pas* was finally got over without accidents—but I shall always recollect that last bit of the descent from Suli as one of the most break-neck samples of imperial Ottoman roads which ever came within my varied experience of Eastern travel.

Near the village of Ghlyky, which seems to have usurped the name of the ancient harbour at the mouth of the river, there are still some scattered ruins of the oracle where the Greeks in the time of Herodotus used to call upon departed spirits, and even to this day the gloomy gorge of the Acheron is associated in the legends of the Greek peasantry with the dark kingdom of the dead. Although it is only a day's ride from Kako-Suli to Parga, we determined to break the journey in order to give our horses a rest after the feats of equilibrium they had been called upon to perform during the last two days; and we took up our quarters in the large Albanian village of Turcopaluro, which is built on a slight eminence out of the marshy plain. Here my faith in my Greek friends of Yanina, who had assured me that south of the Kalamas there were no Albanian communities whom at least the bond of a common tongue did not unite to Greece, was first shaken. My hosts up at Suli had spoken Greek as well as Albanian, and even knew Turkish, probably owing to their intercourse with the garrison of the fort. But here was a village, and a Christian village, where, with the exception of the priest, not a soul either spoke or understood a word of anything but Albanian. Nor, if our Albanian servant could be trusted as an interpreter, had their sympathies any more affinity to Greece than their language. Their village was the property of an Albanian Mussulman Bey, who had never been near the place, and seemed to let his peasantry do very much what they pleased. Most of them were armed; and the pride with which they mentioned that they were allowed to bear arms while the Greek peasantry to the south were not, showed how vastly superior they considered themselves to a race which could perhaps read and write, but could never fight.

The road to Parga led over a succession of low undulating hills covered with the thick-set bushes of the prickly palluria, and there was nothing to relieve its monotony save the flocks and tents of a few Wallach winter settlements, until we reached the coast. Suddenly, as we arrived on the brow of a long ridge somewhat higher than the rest, the Adriatic came into view, its blue waters lapping lazily against the grey cliffs, while the land, sloping gently down towards them, was covered with well-tilled fields and groves of dark olive-trees. Close to a ruined old watch-tower stood a handsome new *chiftlik* belonging to Abeddin Pasha, the ex-Turkish Minister for Foreign Affairs, who, together with his brothers, owns a large amount of property in the Albanian districts of Epirus.

It is near this point of the Albanian coast that has been laid the scene of two ancient legends, between which, however dissimilar their incidents, it is difficult not to imagine some curious connection. The first is not unknown to English readers, and I cannot improve on the quaint version of it given by the old annotator on Spenser's 'Pastoral in May.' "Here, about the time that our Lôrd suffered His most bitter passion, certain persons sailing from Italy to Cyprus at night heard a voice calling aloud, Thamus! Thamus! who, giving ear to the cry (for he was pilot of the ship), was bidden, when he came near to Pelodes, to tell that the great god Pan was dead, which he doubting to do, yet for that when he came to Pelodes, there was such a calm of wind that the ship stood still on the sea unmoored, he was forced to cry aloud that Pan was dead, wherewithal there was such piteous outcries and dreadful shrieking as hath not been the like. By which Pan of some is understood the great Sathanas, whose kingdom was at that time by Christ conquered, and the gates of hell broken up; for at that time all the oracles surceased, and enchanted spirits that were wont to delude the people henceforth held their peace." The other legend was recounted to me by a learned Mussulman of Yanina, who believed that it had been adopted by the Moslems from an earlier Greek tradition, though its memory appears to have entirely died out among the Christian peasantry of the neighbourhood. "After Judas had sold his Master to the Jews for thirty pieces of silver, the traitor fled in terror of revenge to the seaport of Jaffa, and embarked on the first ship which was about to sail. Now Judas was not aware that the crime to which he had ministered was never consummated, for that God blinded the Jews and caused another man to be crucified in the place of His Prophet Jesus; and throughout the journey he was oppressed with remorse, until one night, as they neared the coast of Greece, where the rivers flowing out from the infernal regions pour themselves into the sea, he threw himself overboard with a loud shout, saying that he had killed the Lord his God, and that the spirits of hell were awaiting him. But death being too mild a punishment for the heinousness of his offence, he was borne up by the hands of spirits and carried to the mainland, where he was abandoned to his fate, condemned to roam for all eternity in fruitless search after the kingdom of the dead."

Gradually, as we approach Parga, the cultivation grew more and more luxuriant; the gnarled olive-trees, symmetrically planted on carefully-laid-out terraces, formed a thick forest on either side of the path; and when we debouched out of it on to the beach, we found ourselves in a small, almost land-locked bay, with the town of Parga immediately facing us. A rocky island, which bears the ruins of a convent and of a fort, encloses the bay to the south. At the northern extremity of its outlet a lofty promontory, crowned by the Turkish citadel, projects into the sea; while the town, backed on all sides by green orchards, runs up the slopes of the amphitheatre of wooded hills which protect it from the chill blasts of the north and east. Parga has had a checkered history. It was the last of the Venetian settlements on the coast of Albania, and passed, with the other possessions of the republic, into the hands of France when Napoleon discrowned the Queen of the Seas. An inscription on the guard-house of the small island-fort—"Pour la defence (*sic*) de la patrie, 1808"—still recalls the memory of the French garrison which at that time held it. In 1814 the British squadron which was blockading Corfu landed an expedition on the coast, which, with the connivance of the inhabitants, secured the fortress by a bold *coup de main.* But it was fated to be an evil hour for the Pargiotes when the

British flag replaced the tricolor on the battlements of their citadel. While diplomacy was busy in remodelling the map of Europe at the Congress of Vienna, the little seaport on the coast of Albania seems to have escaped the memory of the British statesmen; and when Turkey, invoking the silence of treaties with regard to Parga, claimed its restitution as forming part of the mainland possessions which were to be restored to her, King George's Government could find no answer but compliance; and the Governor of the Ionian Islands received orders to withdraw the garrison, and hand over the ill-fated town to its most inveterate foe, Ali Pasha, the tyrant of Yanina and the destroyer of Suli. No words of mine can render more vividly than Sir A. Alison's account, the pathetic circumstances which accompanied its surrender.

"When it was rumoured, after the Treaty of 1815, that Parga was to be ceded to the Turks, the inhabitants testified the utmost alarm, and, justly apprehensive of the consequences of being ceded to the Sultan's dreaded satrap, they made an urgent application to the British officer in command of the garrison, who, by order of Sir Thomas Maitland, then Governor of the Ionian Islands, returned an answer in which he pledged himself that the place should not be yielded until the property of those who might choose to emigrate should be paid for, and they themselves be transported to the Ionian Islands. An estimate was then made out of the property of the inhabitants, which was found to amount in value to nearly £500,000, and the inhabitants were individually brought up before the Governor and interrogated whether they would remain or emigrate; but they unanimously returned for answer that 'they were resolved to abandon their country rather than stay in it with dishonour, and that they would disinter and carry with them the bones of their forefathers.' Commissioners had been appointed to fix the amount of the compensation which was to be awarded by the Turkish Government to such of the inhabitants of Parga as chose to emigrate; but they, as might have been expected, differed widely as to its amount, and in the end not more than a third of the real value was awarded. Meanwhile, Ali Pasha, little accustomed to have his demands thwarted, and impatient of delay, repeatedly threatened to assault the town and reunite it to his pashalic without paying one farthing of the stipulated indemnity. At length, in June 1819, the compensation was fixed at £142,425, and Sir Frederic Adam gave notice to the inhabitants that he was ready to provide for their embarkation. The scene which ensued was of the most heartrending description, and forcibly recalled the corresponding events in ancient times, of which the genius of antiquity has left such moving pictures. As soon as the notice was given, every family marched solemnly out of its dwelling without tears or lamentation; and the men, preceded by their priests and followed by their sons, proceeded to the sepulchres of their fathers, and silently unearthed and collected their remains, which they put upon a huge pile of wood which they had previously collected in front of one of their churches. They then took their arms in their hands, and, setting fire to the pile, stood motionless and silent around it till the whole was consumed. During this melancholy ceremony, some of Ali's troops, impatient for possession, approached the gates of the town, upon which a deputation of the citizens was sent to inform the English governor that if a single infidel was admitted before the remains of their ancestors were secured from profanation, and themselves with their families safely embarked, they would instantly put to death their wives and children, and die with their arms in their hands, after having taken a bloody revenge on those who had bought and sold their country. The remonstrance was successful; the march of the Mussulmans was arrested, the pile burnt out, and the people embarked in silence with their wives and children."[1]

For many years after these tragical events, Parga was a wilderness — even its conquerors were afraid to settle in its desolate houses, haunted by the ghosts of the past. But when the war of independence drove many of the Mussulman inhabitants out of the Morea, the Turkish Government gave the refugees grants of land about Parga, and gradually the new colonists were joined by people from the interior who were attracted by the fertility of

[1] Alison's History of Europe, 1815-1852.

the soil, and some even of the families who had emigrated to the Ionian Islands were at last induced to return to their former homes. Nowadays there are scarcely any traces left of that romantic episode in the history of Parga, save the English broad-arrow on some of the guns of the citadel. The population, which is composed in almost equal parts of Mussulmans and Christians, lives in good-tempered amity. Trade is prosperous; and though nature has denied Parga the one thing which would have made it entirely blessed—viz., a harbour—small Greek and Italian craft run into its sheltered bay, and convey the fragrant fruits of its lemon and orange groves, and its plenteous cargoes of olives, to Paxos, whence the Austrian steamers carry them to Trieste and Brindisi, and the markets of central and southern Europe. Among the produce of Parga there is one speciality which is held in high estimation by the Polish Jews: it is a species of sweet lemon which is prepared and candied in a peculiar way for their markets; and so great is the demand for it at the time of the Passover feasts, that, when the crop is deficient, it commands the most fabulous prices. Its flavour is perhaps more delicate than that of other species, but I confess I was unable to discover what peculiar excellency it possessed.

The castle is a picturesque Venetian fortress with straggling walls running up the hillside, but can scarcely possess at the present day any strategical value. I was, however, able to judge of it only from the outside, as here again the officer in command met me with the same *non possumus* as at Suli. By a strange coincidence, he too had received a few days before stringent orders from Yanina not to allow any travellers within the walls of his stronghold.

# CHAPTER XIII.

## AMONG THE ALBANIANS OF SOUTHERN EPIRUS.

From Parga I determined to visit the mountain district of the Tchamouria — a region almost exclusively held by Mussulman Albanians, and where the information I had received at Yanina left me little doubt that a strong and genuine agitation was in progress against Greek annexation. Bidding farewell to the bright little town of Parga, we turned once more into the thick groves of olive-trees which we had crossed on our way from Suli; but about three-quarters of an hour from the shore, we struck northwards over the hills, and soon found ourselves once more among bleak and rugged uplands. A narrow defile between limestone rocks leads into the plain of Margariti, a fertile table-land locked in on all sides by hills, which is the heart of the

Tchamouria, though on most maps the latter name is given to a district south of the Acheron, between that river and Prevesa. The mistake is much the same as if geographers applied the name of British Isles exclusively to the Channel Islands. The limits of the Albanian-speaking districts of Epirus south of the Kalamas may be roughly defined as follows: Starting from the Kalamas near the sharp bend which that river takes to the north at the foot of Mount Lubinitza, they follow the crest of the amphitheatrical range of Suli as far as the gorge of the Acheron. In that neighbourhood, probably owing to the influence which the Suliote tribe at one time enjoyed, they drop over to the east into the valley of the Luro, and follow its basin as far as the peninsula on which Prevesa is situated, where the Greek element resumes its preponderancy. Within these outer limits of the Albanian tongue the Greek element is not unrepresented, and in some places, as about Paramythia, for instance, it predominates; but, on the whole, the above-defined region may be looked upon as essentially Albanian. In this, again, there is an inner triangle which is purely Albanian—viz., that which lies between the sea and the Kalamas on the one hand, and the waters of the Vuvo on the other. With the exception of Parga and one or two small hamlets along the shore, and a few Greek *chiftgis* on Albanian estates, the inhabitants of this country are pure Tchamis—a name which, notwithstanding Von Hahn's more elaborate interpretation, I am inclined to derive simply from the ancient appellation of the Kalamas, the Thyamis, on both banks of which stream the Albanian tribe of the Tchamis, itself a subdivision of the Tosks, has been settled from times immemorial. From the mountain fastnesses which enclose this inner triangle, the Tchamis spread out and extended their influence east and south; and the name of Tchamouria, which is especially applied to the southernmost Albanian settlements in Epirus, was probably given to that district by themselves as an emphatic monument of their supremacy; but it cannot belong less rightfully to the centre, where they hold undivided sway.

The slopes of the hills which surrounded the oblong plateau, at the southern extremity of which we were now standing, were dotted about with numerous villages, and on the eastern side the slender minarets of Margariti were conspicuous amidst masses of green foliage and the square flat roofs of massive fortress-like houses. Margariti is the capital of the district, though it is scarcely larger than Mazaraki, the highest houses of which could be descried above an intervening spur at the northern end of the valley. An hour's ride along the foot of the eastern hills brought us right up to Margariti. Our visit proved to be singularly well-timed, for —— Pasha, the Governor of ——, himself one of the most active promoters of the League in Epirus, had arrived on the preceding day to superintend the enrolment of the Redifs, or first ban of the reserves, who had just been called under arms. Moreover, it was the first day of Baïram, and the whole population had turned out in holiday attire to celebrate this doubly auspicious occasion. Groups of tall, handsome mountaineers, decked out in clean *fustanellas* and gorgeous embroidered jackets, were loitering about the streets; and their proud bearing, more even than the weapons which bristled in their belts, showed clearly that we were among an eminently warlike race. The arrival of two foreigners only added to the excitement which was visible in every face; and we were soon surrounded by self-constituted guides, who volunteered to lead us to the Pasha's residence. As we were already old acquaintances, our reception was most cordial. Quarters were soon found for us in one of the few Christian houses of the place; and we received a pressing invitation to be present at a great demonstration which was to come off on the following day at Mazaraki.

While our Christian host was preparing the evening banquet, we strolled about the quaint little town, which counts nearly 4000 inhabitants, of which only a small fraction are Christians. It is built in an amphitheatre formed by two spurs which project out of the main ridge westwards into the plain. The principal portion of the town—*i.e.*, the bazaars and the houses of the lower classes—lies in the hollow, while the residences of the wealthier citizens and beys occupy the hill-slopes. The summit of the steep bluff to the south is crowned by an imposing castle, built by Ali Pasha to check the bold mountaineers, whom he never succeeded in completely subduing. The grey walls of the solid stone houses, square and massive, would form a somewhat monotonous *ensemble*, were they not relieved by the luxuriant and varied tints of poplar, chestnut, and walnut trees, and the picturesque domes and minarets of the mosques. The bazaars were closed on account of the festival; but as there is no local industry in this district, they were scarcely likely to present any features of special interest, for the wants of the people whom they supply are of the simplest order. When we returned to our quarters, there was, however, alas! no supper ready. An Eastern host would rather keep his guests starving than serve them a modest but hasty meal, which would not, in his opinion, do credit to his hospitality. But coffee and cigarettes, and cigarettes and *raki*, combined with the garrulousness of Mr Triantaphilos himself, helped to while away the time. Although a Christian and a Greek, and in a town where the Mussulman Albanian element was supreme, Mr Triantaphilos seemed to have little to complain of, either for himself or his community. His leanings towards Hellenism were eminently Platonic, his chief anxiety appearing to be lest a war should invalidate his contracts with the Government; while, on the other hand, he seemed convinced that the Greeks could never make him more than he already was—viz., a member of the Municipal Council,—a proud position, upon which he did not fail to lay proper stress. At length the repast appeared, and it certainly justified the long waiting. Besides the inevitable sheep, fowls, turkeys, geese passed successively through the indefatigable carver's fingers; I say fingers advisedly—for though knives and forks were provided for the European guests, mine host evidently never dreamed that they could be used for the purpose of carving.

The first part of the programme the next day was to attend the Pasha's levee. In the corner of a large and lofty room, the only adornment of which consisted in the brilliant carpets which lined three-quarters of the floor, and the delicate wood-work of the panels and painted ceiling, his Excellency squatted *more Turcorum* on a heap of rugs. We took up our places beside him, and soon saw apparently the whole male population of Margariti defile before us,—Imaums and Kadis with green turbans and flowing *kaftans*,

Ulemas with white turbans, and Albanians of every rank and class, differentiated only by the fulness of their starched petticoats, the brilliancy of their jackets, and the gorgeousness of their belts, into which daggers and pistols were stuck with indiscriminate profusion. But to detect the rank of every visitor, it was only necessary to watch the Pasha. Himself an Albanian of these parts, and the owner of large estates in the Tchamouria, he was evidently versed in the jealous rules of local etiquette, and according to the standing of every guest he modified his greeting, now saluting them only with a faint motion of the hand to heart, lips, and forehead, now rising on one knee to perform the salutation, now on both, and now again standing full upright to welcome some personage of transcendent distinction. The code of etiquette is rigid in all parts of the East, but nowhere more so than in Albania. The reason for this is not far to seek —it lies in the fundamental constitution of Albanian society. It is a society made up of castes. In Northern Albania, where clan distinctions are more strongly marked, especially among the Christian tribes, the distinction of castes has been slightly overshadowed by the more obvious features of the clan system; but it exists throughout Albania, and is pre-eminent among the Tchamis. The highest of these castes is that of the warriors, who are at the same time the landowners; next to them rank the artisans and traders; then the shepherds; and lastly the husbandmen. At the first blush it might appear that this classification was arbitrary, as there are many among the smaller arm-bearing landowners who till their own soil, while others combine trading with farming; and among the shepherds and the husbandmen the larger majority also carry arms, and are only too prone to lay aside the crook or the plough in favour of the rifle or the lance. But this objection is merely superficial. Every man's caste is determined by his chief avocation. The warrior-landlord does not cease to be the warrior-landlord because he takes a share in the labours of his peasantry or sells the produce of his estates; but the artisan and the shepherd and the husbandman do not rise out of their respective castes because they are always ready to follow their leaders on the war-path. The chief proofs that this rigid distinction exists are furnished by the two undisputed facts that professions in Albania are hereditary, and that marriage is limited almost exclusively to between families of the same profession. The landowner's sons inherit his estates; if he have no male issue, his eldest daughter's husband, or, if female issue also fail, then the next male of kin. The artisan's son is an artisan, and so on. No Albanian would ever think of changing his own calling or of bringing his children up to another calling than his own. In like manner with regard to marriages. A squire's son marries a squire's daughter; he would no more wed an artisan's, than an artisan would give his daughter to a shepherd or a tiller of the soil. The line is sharply defined; none would ever dream of overstepping it. It has been thus drawn for ages, and in the Albanian's eyes it requires no other sanction. These four castes may crystallise into different social masses according to local custom. In the north and centre, especially among the Christian Albanians, the clan formation prevails, as we have already said; in other parts—about Scutari, Elbassan, and Berat, for instance—small confederacies have formed themselves around the large towns which rule them through their landed aristocracy; in the south, with which we are more especially concerned, the Albanians group themselves around the leading families, their relations with those families resembling sometimes those of feudal retainers, sometimes those of Roman clients. But whatever the social agglomerate may be, those four distinct factors invariably enter into its composition. The same hard and fast laws which govern the relations of the castes with each other within the social group which they compose, equally govern their relations with the corresponding castes of other groups. Though, as a rule, there is little intimate intercourse between the different groups, still there is nothing to prevent a shepherd of the north marrying a shepherd's daughter of the south; but it would be just as unbecoming for a squire from Scutari to wed an artisan's daughter from the Tchamouria, as to contract a *mésalliance* in his own native town. It is on the universality of these caste distinctions throughout Albania, that the leaders of the Albanian movement found their hopes of welding the nation into a single homogeneous mass. The grouping of clans, confederacies, and families (families in the more comprehensive sense of the word, including retainers and clients), is, according to them, a mere adventitious and secondary formation. The elementary formation upon which the whole social structure rests, is the division of castes, and that is common to all Albania.

The levee ended, we mounted our horses and rode over the plain, rounding the spur which is projected into it from the western ridge of hills, towards Mazaraki, where a great gathering of the Tchamis had been convoked. The careful cultivation of the soil, and the number of villages clustering on the slopes of the mountains, showed how populous this district is. Most of the notables of Margariti accompanied the Pasha, mounted on every variety of horse and mule; even the modest donkey was not unrepresented in the cavalcade, which glittered in the sunshine with the rainbow tints of oriental dress. Some hundred bold mountaineers on foot careered merrily in front of us, uttering wild war-whoops, and discharging in the air their bell-mouthed pistols and ancient matchlocks with interminable barrels. After a couple of hours' smart riding, we reached the meadow where the demonstration was to take place. It was indeed a striking spectacle. Not less than 2000 Albanians had responded to the call. As we took up our places on the carpets which had been spread for the Pasha's party under the shadowing branches of an olive-grove which fringed two sides of the meadow, the whole assemblage rose to their feet and saluted their leader with a long low shout. The scene was one of surpassing solemnity. The picturesque groups of mountaineers, their red caps and white *fustanellas* contrasting with the green meadow, their strange uncouth weapons gleaming in the sunshine, the grey houses of Mazaraki nestling among trees on the slopes of the western hills, the wooded heights behind them, and in the background distant ranges of blue mountains rising tier above tier against the brilliant sky. Our host rose in response to the acclamations of the meeting, no longer the stolid Turkish Pasha, but the enthusiastic patriot; and as the winged words flew from his lips, he seemed to breathe his very soul into the assembled multitude, until, when the closing sentence fell upon them, every man took it up and repeated it with the full power of two thousand healthy lungs, "Long live the autonomy of Albania!" Then came a striking incident. Among the mountaineers gathered on the meadow, there were many who belonged to villages and families between whom there existed ancient feuds. Solemnly these bitter foes of yesterday walked up towards the Pasha and bound themselves over, with the kiss of peace, to abjure their quarrels for evermore, in view of the national peril. Such a thing has never been seen, the Pasha told me, within the memory of man. "And look there," he added, pointing to a small knot of mountaineers—"those are Christians. Hitherto they have always held aloof from their Mussulman neighbours, but to-day they have come unsolicited to join in the common demonstration. There are not many Christians in the Tchamouria, but when arms are distributed, I shall give them with the same confidence to Boutro (Peter) as to Ahmed. For, mind you, it is not with those old matchlocks that they can fight the Greeks, however brave and fearless they may be. The gist of my speech was that the nation was in danger, and that I had come to supply them with arms and organise their resistance, that they might defend the liberties of a united and independent Albania."

In the evening we were the Pasha's guests, and after supper he entered at great length into the scope and signification of the Albanian movement, and fully confirmed what I had gathered of its bearing from the members of the League in Yanina. In Europe it was the habit to denounce the League as the mere docile instrument of Turkish hostility towards Greece. The Albanians, however, were neither hostile to Greece nor slaves to Turkey. They did not forget that many of their own kith and kin had bled for Greece during the war of independence, and that at the present day the King of Greece counted 200,000 Albanians among his subjects. Nor did even the Mussulmans, who were bound to Turkey by the bond of a common religion, forget that it was only by dint of fighting that their fathers and forefathers had conquered from the Sultans the privileges which they were still allowed to enjoy. They claimed at the present day, when the voice of every nationality in the East was being heard, that theirs also should be listened to. They were determined not to allow themselves to be absorbed either by Greece or by Turkey. The watchword of the hour was the provinces of the Balkans for the nationalities of the Balkans, and the Albanians had taken it up in order to assert the right of the Albanians to Albania. If Greece refused to recognise that right, so much the worse for Greece. If Turkey refused to acknowledge it, so much the worse for Turkey. Nature had been a stern mother to Albania, but she had at least given the Albanians stout hearts and strong sinews to fight their own battles, and those were in future the only battles which they would fight. The Tchamis south of the Kalamas were a small and insignificant tribe, four or five times outnumbered by the Greeks of Southern Epirus, and it might be contended that so trifling a minority had no claim to a hearing in the settlement of the destinies of that province. But they had the most sacred of duties to fulfil. They were placed as sentinels on the southern outposts of Albania, and they were bound to defend with the last drop of their blood the post of honour and of peril which nature had intrusted to them. They had no rancorous animosity towards Hellenism. On the contrary, they had in many ways felt its influence for good. They owed to it much of the little education they possessed. But they were not disposed to sacrifice their birthright to the ambition of Athenian politicians. Let Greece treat with them on a footing of equality, and there was nothing to prevent an intimate union between the two nations. But they were not prepared to exchange the rule of Constantinople for that of Athens, and the measure of liberty which they enjoyed under the Sultan's sovereignty for the incubus of Greek centralisation. To prevent such a consummation, they relied not only on their own arms, and on those of the Albanians north of the Kalamas whose essential interests were equally at stake with their own, but on the justice of Europe. It was no religious fanaticism against their Christian neighbours which inspired the movement, but their strong national settlement, which, in Albania alone, almost of all countries in the East, was more powerful than sectarian prejudices. The Tchamis had the reputation of being the most fanatical among Moslem Albanians. Religion sat, perhaps, more lightly on the Mussulmans of northern and central Albania; but the Tchamis were Albanians first, and only in the second place followers of the Prophet. Albania was determined to vindicate her right to independence against all comers, and no religious considerations would restrain the Tchamis from joining in the struggle for the common cause, whatever faith its adversaries might profess. The European statesmen of all parties have repeatedly stated that their policy is directed solely towards the protection of the populations of the East without reference to creed distinctions. How can they refuse to listen to the Albanians, who claim a hearing, not as Christians or as Mussulmans, but on the ground of a nationality common to both?

These are the leading ideas which appear to inspire the present movement, and which I have heard out of every Albanian's mouth, now couched in stronger, and now in more guarded terms, according to the speaker's position or character. Its prospects, and the turn which it is likely to take, depend upon the attitude of the Turkish and Greek Governments respectively towards it. The Greeks do not yet appear to have appreciated the immense value of the support which they might obtain from Albania by the simple recognition of its rights, and by renouncing vain dreams of annexation. On the other hand, the Porte scarcely seems to have realised on what a slender basis its rule in Albania rests. To encourage and strengthen the League for momentary objects, it has gradually withdrawn its military and administrative hold on the Albanian provinces of the empire. It has handed over all the public offices and administrative posts to Albanians; and the troops who garrison the country, with the exception of those under Dervish Pasha at Scutari, are Albanian Nizams and Redifs. On the day when Greek advances or Turkish stubbornness induce the League to give the word of command, the officials and the troops throughout Albania will throw off the *Stamboulina* and the Turkish uniform, and appear as what they are, notwithstanding the disguise they wear to-day—the servants not of the Porte, but of the League.

The slanting rays of the sun had just reached the straggling houses of Mazaraki, embowered in pleasant masses of green foliage, as we took leave on the following morning of the Albanian Pasha, and turned our horses' heads towards the northern mouth of the valley. It was a delightful ride in the crisp autumnal morning, up hill and down dale, past cheerful villages and luxuriant plantations of olive and of walnut, to the small port of Gomenitza. The extensive ruins of a Venetian fort bear witness to its past. As to its present not much can be said, except that nature has favoured it with a most lovely position. It lies at the head of a deep bay, enclosed by green and wooded heights, above which tower the mountains of Albania; and on the western horizon, out of the blue waters of the Adriatic, the cliffs and hills of Corfu rise in bold relief against the sky. Here I chartered a small barque to convey me across to Corfu; and as the coasts of Albania slowly receded in the gloaming, it was with a heavy heart I bade farewell to a country whose inhabitants had won my regard more quickly by their manly bearing, their brave hearts, their ready wits, and their straightforward speech, than any other race I had met with in the East since I left the Druses of the Syrian Hauran.

# CHAPTER XIV.

## THE ALBANIAN LEAGUE.

Less than a century ago, Gibbon, writing of Albania, was still fain to confess that though within sight of Italy, it was less known than the interior of America. Since then it has been explored by numerous travellers who have described the beauty of its scenery, the picturesqueness of its towns, the manners and customs of its inhabitants, and have studied their quaint folk-lore and their strange unwritten tongue. But though all agree to recognise in the Albanians the descendants of a most ancient race, the mystery of their origin has never yet been satisfactorily unravelled. Nor is this the place to enter at length into the question, to which a valuable contribution has lately been made by a French professor, M. Benloew, in an essay entitled, 'La Grèce avant les Grecs.' Though there are many points in his work which are evidently strained to support a theory, his sometimes fantastic conjectures do not detract from the value of his more serious arguments; and the new light which he often throws with the help of the Albanian language on the nominology of ancient Greece, as well as the skill with which he traces the influence of Asiatic traditions on Greek mythology, substantially support his contention that the Albanians of to-day are the direct descendants of the first immigrants from Asia Minor into the south-eastern peninsula of Europe, of the Pelasgi and Lelegi who were afterwards displaced by the hardier Aryans known to history under the name of Greeks, and compelled to seek refuge in the wild mountain-ranges where they have preserved into modern times the type, the language, and many even of the customs of their ancestors. In corroboration of this theory, I may mention that an Austrian craniologist whom I met on my way back to Constantinople, told me it was only in Albania that he found skulls similar to the peculiar formation of those discovered in the old Illyrian tumuli of Dalmatia.

Whether the Pelasgic solution of the problem be finally adopted, or whether preference be given to the Semitic solution, or whether their own pretensions to autocthony obtain recognition, the exact origin of the Albanians does not seriously affect the future of the Albanian question. To deal with the latter, it is sufficient to know that the race which occupies at the present day the mountainous regions of the western portion of the Balkan peninsula is one of exceeding antiquity, which has preserved through centuries, and amid the floods of successive barbarian invasions, not only the purity of its lineage and language, but the continuity of its national traditions and institutions; and though many pages in the history of Albania are still blank or obscure, enough of it is known to point the suicidal folly of the Porte in attempting to kindle once more into flame for its own selfish purposes the fiery spirit of Albanian independence which it had taken so many centuries to quench. It has already sadly burnt its fingers in the undertaking, and worse may yet come. But then why does the Porte never recollect that States, any more than children, should be allowed to play with fire?

The history of Albania since the advent of Ottoman rule in Europe until the beginning of the century, is merely the record of perpetual struggles with varying issues between the hardy mountaineers and the satraps sent to rule them from Constantinople; and the power which the latter enjoyed, depended more upon their success in applying amid the local jealousies of contending parties the old maxim, *Divide et impera*, than on the respect or awe which the central Government could inspire in these remote regions. Ali Pasha, the tyrant of Yanina, was the first who attempted with an iron hand to weld into one homogeneous mass the *disjecta membra* of Albania. The success, however partial and ephemeral, which attended his efforts, was due doubtless, in the first place, to his monstrous genius, for which no undertakings were too laborious and no crimes too atrocious; but it may well be doubted whether even he would have succeeded in breaking the power of the feudal aristocracy and coercing into subjection the stubborn necks of the Skipetars, had he not possessed in their eyes the supreme qualification of a national ruler, in that he himself was born an Albanian. Thus it happened that when his power at last roused the jealousy of Constantinople, and the octogenarian tyrant fell pierced by the bullets of the Turkish soldiery, the Porte found itself deceived in the hopes which it had fondly formed of substituting its own yoke for that which it had indirectly helped the Albanians to throw off. It was not till ten years later—not till after the treacherous banquet in the tented field of Monastir—that Albanian liberty lay foully murdered at the Sultan's feet. Even then certain concessions were still granted to the more powerful of the tribes, and local privileges confirmed or accorded in order to allay their resentment. But in the main, Albania was henceforth assimilated to the rest of the empire: Turkish pashas occupied the *konaks* of her cities; Turkish soldiers mounted guard over her strongholds, while her own sons were drafted into regiments far away in Syria or in Asia Minor; Turkish judges rendered Turkish justice in her law-courts; and Turkish tax-collectors robbed her peasantry to build palaces on the Bosphorus. Now and then Albania writhed under the lash; but the harsh discipline seemed to have cowed her proud spirit, while the local feuds and tribal jealousies which the Turkish authorities unceasingly laboured to develop and intensify, appeared to incapacitate her for all united action.

In stating that the present agitation in Albania was not spontaneous in its origin, I do not in the least intend to disparage its value or question its actual genuineness. The rapidity and vigour of its growth sufficiently testify that the germs of the national movement were embedded in the heart of every Albanian; and if they were first quickened into life by the preachings and exhortations of the Porte, it was the virtues and patriotism of the Albanians themselves which ripened and could alone ripen them into maturity. The principle of nationalities had been so often made use of against the Ottoman Government as a pretext for spoliation and foreign interference, that it was only natural for the Porte to seize the first opportunity which presented itself of turning the same weapon against its adversaries, and defeating by the same arguments their demands for further territorial sacrifices. European diplomacy had adopted for its watchword, "The provinces of the Balkan peninsula for the populations of the Balkan peninsula." To plead that some portion of those provinces should be reserved for their Turkish populations was out of the question, for they had been voted an anti-human race, and as such, had forfeited all their rights. But the Albanians had not been outlawed; and though many of them, as Mussulmans, had no doubt incurred the reprobation which enveloped all their co-religionists, there was among them a considerable leaven of Christians who might be expected to deserve at the hands of Christian Europe some measure of that sympathy which was so profusely lavished on Greeks, Bulgars, and Armenians: it might even be hoped that the warmth of that sympathy would be reflected in a lesser degree on their Moslem brethren. The Powers demanded that the national aspirations of the Greeks and Montenegrins should be satisfied by a rectification of frontiers; what more natural than for the Porte to reply that, in regard to such rectifications, account must also be taken of the national aspirations of the Albanians? The retort was just, if founded in fact. The Powers could only meet it by denying that the Albanians had any national aspirations; so the denial was given, and the Porte set to work to prove the *bonâ fides* of its plea. The Albanian agitation dates from the day when this ingenious idea was conceived and adopted at the Palace of Yeldiz. Its growth may be conveniently divided into three periods.

The first period is that in which it was transplanted from the shores of the Bosphorus to the mountains of Albania, and nurtured under the watchful care of the Ottoman authorities. The Central Committee of the League, which had been founded at Stamboul, established branches in all the principal towns of Albania, and the adhesion of the Albanian chieftains was canvassed for by the Turkish governors. Two Albanian chieftains were sent as ambassadors to the European Courts to announce the birth of the Albanian question, and emissaries were despatched to preach the new message in the remotest corners of the Albanian highlands. The seed fell upon good soil, grew up, and bore fruit a hundred-fold. In the course of a year the idea conceived at Yeldiz had become a living reality, for which brave men were willing to lay down their lives. When the Christian tribes of Northern Albania showed themselves on the bridge of the Zem, and, by opposing the advance of the Montenegrins, frustrated the execution of the Corti Convention, it became evident that a new factor had been introduced into the problem, the strength and value of which were still uncertain, but which would at any rate necessitate a revision of all previous calculations. The Sultan saw only the successful blow which his new weapon had enabled him to aim at European diplomacy. He had not yet felt the keenness of its double edge. Saïd Pasha, more perspicacious than his master, advised him to be cautious in the handling of it. But the Sultan was not in the mood to be cautious. The weapon had been cast of Albanian metal: to whom could it better be intrusted than to an Albanian? Saïd was dismissed, and Abeddin called to power.

The Albanian question now enters on its second period. Abeddin Pasha, though holding office under the nominal leadership of Kadri Pasha, had virtual control over all the departments of State, so far at least as such control is possible where the controllers themselves are little more than puppets in the sovereign's hands. He at least in one point faithfully reflected the Sultan's wishes (and that is in Turkey the secret of power)—viz., that the hands of the League must be strengthened. And he set to work to strengthen them with characteristic energy. Turkish governors and Turkish officials throughout Albania were superseded by native governors and native officials; the law-courts were subjected to a process of *épuration* familiar to our neighbours across the Channel, and peopled with servants of the League; the army stationed in Albania was reorganised—Turkish regiments were replaced by Albanian regiments, and Albanian officers were appointed to all the highest commands. In the course of three months the hands of the League had been strengthened to such a degree that it was ready at any moment to shake off the tutelage of the Porte and take up its ground as an independent Power. The Sultan began to feel the double edge of the weapon which he had forged. Under the threat of the naval demonstration he attempted to replace it in its sheath: but he was no longer able to handle it; it had outgrown its sheath; it was informed with life; its point was turned against his own breast. Saïd, who with the looks had perhaps also the cunning of a wizard, might yet be able to turn its edge. An imperial firman announced the dismissal of Abeddin and the return of Saïd to office.

But the Albanian movement had attained to its majority. It was ready to enter into a new phase. The recall of Ali Riza Pasha from Scutari, and the appointment of Dervish Pasha in his stead, only served to precipitate events. The surrender of Dulcigno and the collapse of the vaunted Albanian resistance seemed for a moment to justify the sceptics who had persistently derided the movement as a mere farce prompted from Yeldiz Kiosk. But I fancy other considerations besides Dervish Pasha's battalions and the international fleet operated to reconcile the League to the cession of the small fishing-town which it had so often threatened to destroy rather than give up. In order to stimulate the hostility of the Albanians so long as it was the policy of the Porte to use them as a weapon against Europe, the emissaries of the Turkish Government had never failed to describe the demand for the surrender of Dulcigno as an act of special malice directed solely against the Skipetars, and dictated by a spirit of fanatical hostility towards their nation; and whenever the Skipetars alluded to the question of their autonomy, it was met by expressions of deep regret that, notwithstanding the desire entertained in the highest quarters to meet the wishes of his Majesty's devoted subjects, it was inexpedient to still further provoke the Powers by a measure which would be construed by European prejudice into an act of overt defiance. When it became known through the public press, through parliamentary blue-books and various other channels, that a proposal for granting to Albania that very autonomy which it had so often claimed and which the Porte had so often refused, had been made in the International Commission at Constantinople by the representative of the Power which had always been denounced as the deadliest foe of the Albanians, the latter naturally began to ask themselves, "*Qui trompe-t-on ici?*" And when it further became known that the proposal, supported by Abeddin, had been negatived by Saïd, who had just returned to power as Grand Vizier, the reaction produced by their revelations was immediate, and induced the League to at once modify his policy. If the Powers were inclined not to crush, but to befriend Albania, what was the use of provoking their resentment by useless resistance? The surrender of Dulcigno was agreed to as a concession, not to the Porte, but to Europe. The hostility of the Albanians was thus deflected into a new channel, and betrayed itself in an attitude of growing defiance towards the authority of the Sultan. Before finally throwing off its allegiance, the League decided to make a supreme effort at Constantinople to obtain the fulfilment of the promises which had hitherto been made only to be broken. A general meeting of the League was held early in November last at Dibra, and attended by deputies from all parts of Albania. A petition, or rather an ultimatum, was drawn up, demanding *for the last time* the recognition of Albanian autonomy. Two delegates were selected to take it to Constantinople and deliver it only into the Sultan's own hands. But before even the return of the delegates—who were not so much as admitted into the imperial presence—announced the failure of their mission, the treacherous arrest by Dervish Pasha of Prink Bib Doda, the prince of the Mirdites, and of Hodo Pasha, the most powerful Mussulman chieftain of Northern Albania, had shown what treatment the Skipetars had to expect at the hands of the Turkish Government, now that the League had outlived the Sultan's liking. War was never openly declared, but it was openly waged. The few Turkish officials remaining in Albania were ignominiously expelled; those freshly appointed by the Porte were turned back by armed force; the Redifs liable to service refused to obey the summons; Ali Pasha of Goussinieh collected thousands of mountaineers around his standard, and turned to good account the long winter months, when the inclemency of the weather alone renders all military operations impossible, in drilling and preparing his forces for the coming struggle. The seat of the Central Committee of the League, which had virtually assumed the reins of Government, was transferred from Prizrend to Dibra — a more central position, where closer and more intimate relations could be established between the Ghegs and the Turks of Southern Albania. Not the least significant incident of the winter was the formation of an Albanian Committee at Athens, through which the League was able to place itself in direct communication with the Hellenic Government. The threat of a Greek invasion has alone deterred the Southern Albanians, and especially the Tchamis, who are more immediately menaced, from openly taking part with their brethren of Central and Northern Albania. But they are not unrepresented in the councils of the League; and should the danger of Greek annexation be averted either by diplomatic negotiations or by a direct arrangement with Greece, we shall probably see the whole Albanian nation, from the shores of the Ambracian Gulf to the borders of Montenegro—one and a half million of souls—united in the vindication of their rights.

Albania now seems to hold her destinies in her own hands: she can shape them according to her own bent. Is the national feeling, the consciousness of a national duty, strong enough to overrule the distracting influences of tribal jealousies and sectarian differences, and combine for a supreme effort all the heterogeneous elements divided by the traditions of secular rivalry? Those who know Albania best are disposed to answer Yes. The pride of creed is in Albania only second to the pride of race. In Albania alone, of all the countries of the East, the first question asked of a man is, not what he is, but who he is. So long as he is an Albanian, it is only a secondary consideration whether he be Christian or Mussulman. Religion is regarded chiefly as a matter of expediency, and there are especially many arguments to show that Islam has never struck deep roots into the soil of Albania. The prevalence of Christian names and Christian customs among Mussulmans; the frequent intermarriages between Christians and

Mussulmans; the existence of numerous communities of crypto-Christians, who follow outwardly the law of the Prophet, and conform secretly to the practices of Christianity; the position of women, which is permanently determined by their birth, whereas in Mohammedan countries maternity can raise the humblest to the highest rank,—these are only a few illustrations of the numerous points in which the creed of Islam has failed to weaken the strength of national traditions. With regard to the petty squabbles and tribal feuds which seem hitherto to have engrossed all the energies of the Albanians, it is asserted that you have only to set before them a higher goal for them to concentrate upon its attainment the courage, the endurance, the devotion, which they have so often wasted to such trifling purpose. I can at any rate but express the hope that this sanguine estimate may be realised. A higher goal has now been set before them—the vindication of their liberty; and certainly, of all the subject races of Turkey, none have better deserved to conquer it, for none have shown themselves more tenacious of its virtues.

# CHAPTER XV.

## BACK TO ISTAMBOUL.

A FEW hours' sail over a narrow channel of the Adriatic brought me, as it were, into another world. From the harbour of Corfu my eyes wandered back to the mountains of Albania, so near and yet so far off, separated only by a few miles of sea, and yet divided off from the scenes around me by the immeasurable gulf of centuries. Everything about me was impressed with the mark of our modern civilisation, from the men-of-war in the roadstead, equipped with all the improved appliances of a deadly science, and the big steamers bearing the spoils of the East to the crowded markets of the West, down to the fussiness of the custom-house officials, the offensive obsequiousness of hotel touts and *ciceroni*, the excited squabbling of coffee-house politicians, the busy bustle of the market-place, and the strutting swagger of dandified Greek officers, with trailing swords and washed-out faces. Yonder, across that narrow strip of blue water, the mountains of Albania loomed out of the mist like mighty barriers, potent to arrest even the march of time. In the peaceful valleys which nestled under their rugged slope generations had succeeded generations, but man remained unchanged, like his solemn surroundings. Now and again the breath of the storms which shattered the thrones and uprooted the civilisations of the world around him had swept over the placid waters of his motionless existence, but they had passed over its surface without stirring its depths: the same unwritten laws governed his social life; the same simple means supplied the same simple wants; the same fierce passions marred the same primitive virtues; if the form of his religion had changed, the old spirit still overshadowed the new dogmas; days, years, centuries had passed, but, for better or for worse, man had learnt and forgotten nothing. Was it for better or for worse? I was just trying to think out that question when a violent thump on my back arrested all further reflections. A young Greek officer, whom I had met a few months previously at Athens, shook me vehemently by the hand, and inquired, with a lisp, where I came from. "From Albania." "Bah! what a country to travel in! Why, it's a land of savages. But wait till we have had it for ten years in our hands. Then come and travel in it. You shall have railways, and theatres, and hotels, and newspapers——" "And a constitution, and party government, and pure elections, and universal shoddy. Thank you for nothing! Civilise Albania by all means, if you yourselves have enough civilisation to spare for the job; but, for God's sake, give me timely warning, that I may move on beforehand to some new land of savages." My friend was puzzled, and I fear somewhat hurt at my outburst. That a man who wore a hat and a suit of dittos could have any sympathy for savages was doubtless an unexpected revelation. He did not understand that one may wear a livery and yet feel inclined at times to revolt against the bondage.

I was inclined to conclude from the constant

tramp of armed men, and the ubiquitous ping-ping of musketry practice, that the Corfuotes had indeed taken in full earnest their mission of civilisation. As, sauntering along the esplanade, I watched old men of sixty and boys of fifteen vying with each other in the performance of the goose-step, I was disposed to admire in them the fortitude of patriots who were ready to brave not only the horrors of the battle-field but the weariness of the drill-yard, in order to make their Greek brethren and Albanian cousins over the water copartners in the blessings of a strongly centralised Government. Alas for the waste of admiration! My heroes were only poor devils, who had been but yesterday brought in from their native villages, where the keen eye and strong hand of the police had ferreted them out of their hiding-places in hay-stacks, cow-sheds, and suchlike, more fortunate, perhaps, but scarcely more worthy of admiration than the six hundred other natives of Corfu who have made good their escape from conscription by flying to Brindisi, Venice, and Trieste. It would seem as if the Greek nomarchs of Corfu had not yet succeeded in establishing the millennium on the ruins of our tyranny and public works. The age of gold and silver has passed away with our rule, and the age of paper money and copper has followed in its stead; but the iron age of Hellenic virtues is not yet.

The eternal beauty of Corfu has been so often celebrated that it is needless to dwell on the twice-told tale. The view of the bay as we steamed out of it at sunset was of more than ordinary loveliness. The sun had already sunk over the hills behind the town, and the faintest of evening mists tempered the hard outlines of the modern houses and the sharp angles of the bastions which rise tier upon tier from the water's edge; the lofty peak of Mount San Salvador, where the people still go up in the summer to the "high place," to worship on the same spot where Nero danced before the altar of Zeus, stood out in dark relief against the crimson sky, while over the glassy waters to the east the Acroceraunian mountains reared up to heaven "their thunder-cliffs of fear." I had taken passage on board one of the Greek steamers which ply between Corfu and the head of the Gulf of Corinth, and I may take this opportunity of saying that for comfort, cleanliness, and punctuality, the Atmoploia Hellenike may safely bear comparison with any other line of steamers in the Levant. By the brilliant light of the full moon I could discern many points which had become familiar to me during my journey through Epirus—the long, low cliffs of Paxos and Antipaxos, the castellated Bay of Parga, the marshy estuary of the Acheron, and far beyond, against the dark horizon, the lightning ridge of Suli.

After stopping once in the middle of the night at Santa Maura, we cast anchor, in the early morning, in the harbour of Zante. The tall square campaniles of its churches, the shady arcades which line its streets, the columned balconies of its massive stone houses, at once mark its long and intimate connection with Venice; while the influence of the Moslem east is apparent in the heavy wooden lattices which shield the windows from the impertinent gaze of strangers. Shut in by precipitous cliffs, in many places rent and torn asunder by earthquakes and landslips, the town stretches along the shore of the semicircular bay to a distance of nearly two miles, and beyond it at either end bright villas gleam amid the dark foliage of orange-groves and olive-trees. A ruinous Venetian castle crowns the hill above the northern extremity of the town, while the eastern end of the bay is guarded by the jagged crest of an extinct volcano. The long quays made bright with the varied colours of flowers and fruits, and the quaint forms and rigging of the native craft crowding behind the mole, make up a picture which is well worthy of its motto : " Zante, Zante, Fior di Levante"—the Water-lily of the East.

From Zante we changed our course to the north-west, and we were soon sailing up the noble entrance of the Corinthian Gulf. To the north the mountains of Ætolia and Acarnania mingled their snows with the fleecy clouds which hung about their summits; and in the midst of the plains and lagoons which stretched away to their feet, a long line of walls and ramparts marked the town of Mesolonghi, where, during the war of independence, the remnant of the Suliotes under the leadership of Mark Botzaris, avenged upon the Turks the destruction of their mountain homes, and shed upon the Greek cause the immortal lustre of their prowess. To the west and south, the jagged ranges of the Peloponnesus tower above the fertile plains of Achaia, Mount Voidhia claiming kingship above them all. Presently Patras, the second town of Greece, comes into sight, prosperous, clean, solidly built, and laid out with painful regularity, a geometrical figure cut out in stone and stucco. Precious hours of daylight were wasted in loading endless boatloads of currants; and as we entered the Straits of Lepanto, where the fleets of Christendom first humbled the pride of the ever-victorious Crescent, the shades of evening were fast closing in over one of the most lovely scenes in Europe. Before morning broke we had reached the modern Corinth. There is nothing in the modern town to arrest attention, and the traveller's eye quickly glances up from its whitewashed hovels and prosaic wine-shops to the isolated rock where the ruins of the Acro Corinthus, tower-capped, "seem the very clouds to kiss." The modern Greeks have often talked of completing the canal which their ancestors planned and even commenced across the isthmus. But in these days of safe investments and speedy returns, the project has not met with much financial support. So the journey has still to be performed by road. An hour suffices to cross the narrow plain, and at Kalamaki another steamer is in readiness to convey the traveller across the Saronic Gulf to the Piræus.

Forty-eight hours later, and exactly eleven weeks after my departure, I landed again at Constantinople. A storm of rain and sleet and bitter squalls from the Black Sea, announced that winter was upon us, and that my journey had been brought to a timely conclusion.

THE END.

PRINTED BY WILLIAM BLACKWOOD AND SONS.

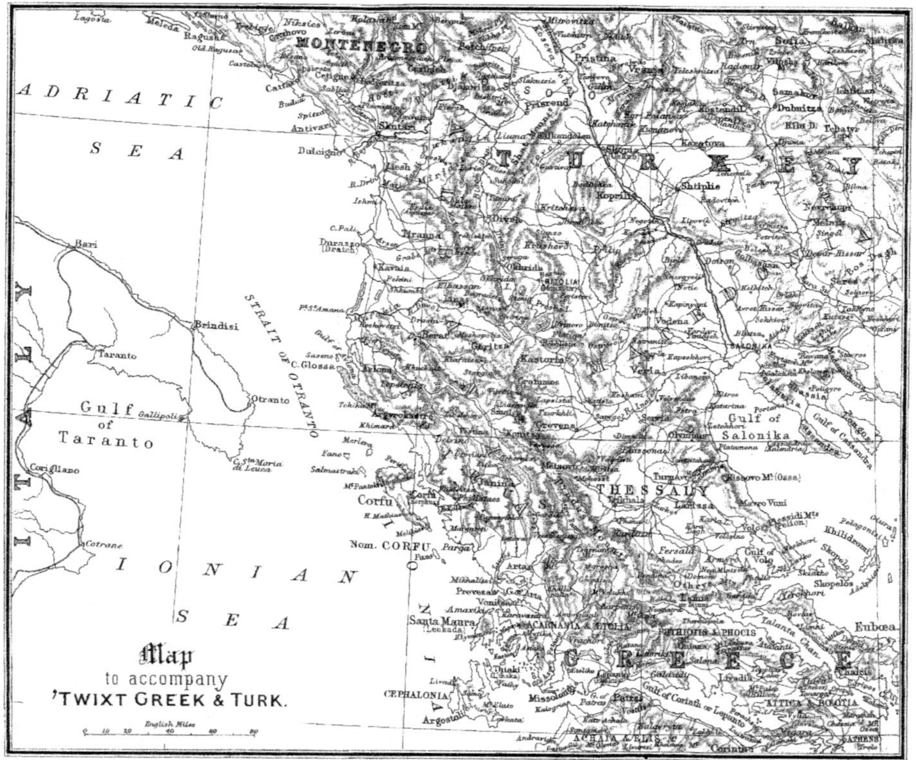


J.Bartholomew Edin^r.

# CATALOGUE

OF

# MESSRS BLACKWOOD & SONS'

# PUBLICATIONS.

# CATALOGUE

OF

# MESSRS BLACKWOOD & SONS'

## *PUBLICATIONS.*

---

ALISON. History of Europe. By Sir ARCHIBALD ALISON, Bart., D.C.L.

1. From the Commencement of the French Revolution to the Battle of Waterloo.
   LIBRARY EDITION, 14 vols., with Portraits. Demy 8vo, £10, 10s.
   ANOTHER EDITION, in 20 vols. crown 8vo, £6.
   PEOPLE'S EDITION, 13 vols. crown 8vo, £2, 11s.
2. Continuation to the Accession of Louis Napoleon.
   LIBRARY EDITION, 8 vols. 8vo, £6, 7s. 6d.
   PEOPLE'S EDITION, 8 vols. crown 8vo, 34s.
3. Epitome of Alison's History of Europe. Twenty-eighth Thousand, 7s. 6d.
4. Atlas to Alison's History of Europe. By A. Keith Johnston.
   LIBRARY EDITION, demy 4to, £3, 3s.
   PEOPLE'S EDITION, 31s. 6d.

——— Life of John Duke of Marlborough. With some Account of his Contemporaries, and of the War of the Succession. Third Edition, 2 vols. 8vo. Portraits and Maps, 30s.

——— Essays: Historical, Political, and Miscellaneous. 3 vols. demy 8vo, 45s.

——— Lives of Lord Castlereagh and Sir Charles Stewart, Second and Third Marquesses of Londonderry. From the Original Papers of the Family. 3 vols. 8vo, £2, 2s.

——— Principles of the Criminal Law of Scotland. 8vo, 18s.

——— Practice of the Criminal Law of Scotland. 8vo, cloth boards, 18s.

——— The Principles of Population, and their Connection with Human Happiness. 2 vols. 8vo, 30s.

ALISON. On the Management of the Poor in Scotland, and its Effects on the Health of the Great Towns. By WILLIAM PULTENEY ALISON, M.D. Crown 8vo, 5s. 6d.

ADAMS. Great Campaigns. A Succinct Account of the Principal Military Operations which have taken place in Europe from 1796 to 1870. By Major C. ADAMS, Professor of Military History at the Staff College. Edited by Captain C. COOPER KING, R.M. Artillery, Instructor of Tactics, Royal Military College. 8vo, with Maps. 16s.

BOSCOBEL TRACTS. Relating to the Escape of Charles the Second after the Battle of Worcester, and his subsequent Adventures. Edited by J. HUGHES, Esq., A.M. A New Edition, with additional Notes and Illustrations, including Communications from the Rev. R. H. BARHAM, Author of the 'Ingoldsby Legends.' 8vo, with Engravings, 16s.

BRACKENBURY. A Narrative of the Ashanti War. Prepared from the official documents, by permission of Major-General Sir Garnet Wolseley, K.C.B., K.C.M.G. By Major H. BRACKENBURY, R.A., Assistant Military Secretary to Sir Garnet Wolseley. With Maps from the latest Surveys made by the Staff of the Expedition. 2 vols. 8vo, 25s.

BROOKE, Life of Sir James, Rajah of Sarāwak. From his Personal Papers and Correspondence. By SPENSER ST JOHN, H.M.'s Minister-Resident and Consul-General Peruvian Republic; formerly Secretary to the Rajah. With Portrait and a Map. Post 8vo, 12s. 6d.

BROUGHAM. Memoirs of the Life and Times of Henry Lord Brougham. Written by HIMSELF. 3 vols. 8vo, £2, 8s. The Volumes are sold separately, price 16s. each.

BROWN. The Forester: A Practical Treatise on the Planting, Rearing, and General Management of Forest-trees. By JAMES BROWN, Wood-Surveyor and Nurseryman. Fifth Edition. Royal 8vo, with Engravings. [*In preparation.*

BROWN. A Manual of Botany, Anatomical and Physiological. For the Use of Students. By ROBERT BROWN, M.A., Ph.D., F.L.S., F.R.G.S. Crown 8vo, with numerous Illustrations, 12s. 6d.

BROWN. Book of the Landed Estate. Containing Directions for the Management and Development of the Resources of Landed Property. By ROBERT E. BROWN, Factor and Estate Agent. Large 8vo, with Illustrations, 21s.

BUCHAN. Introductory Text-Book of Meteorology. By ALEXANDER BUCHAN, M.A., F.R.S.E., Secretary of the Scottish Meteorological Society, &c. Crown 8vo, with 8 Coloured Charts and other Engravings, pp. 218. 4s. 6d.

BURBIDGE. Domestic Floriculture, Window Gardening, and Floral Decorations. Being practical directions for the Propagation, Culture, and Arrangement of Plants and Flowers as Domestic Ornaments. By F. W. BURBIDGE. Second Edition. Crown 8vo, with numerous Illustrations, 7s. 6d.

——— Cultivated Plants: Their Propagation and Improvement. Including Natural and Artificial Hybridisation, Raising from Seed, Cuttings, and Layers, Grafting and Budding, as applied to the Families and Genera in Cultivation. Crown 8vo, with numerous Illustrations, 12s. 6d.

BURN. Handbook of the Mechanical Arts Concerned in the Construction and Arrangement of Dwelling-Houses and other Buildings; with Practical Hints on Road-making and the Enclosing of Land. By ROBERT SCOTT BURN, Engineer. Second Edition. Crown 8vo, 6s. 6d.

BUTE. The Roman Breviary: Reformed by Order of the Holy Œcumenical Council of Trent; Published by Order of Pope St Pius V.; and Revised by Clement VIII. and Urban VIII.; together with the Offices since granted. Translated out of Latin into English by JOHN, Marquess of Bute, K.T. In 2 vols. crown 8vo, cloth boards, edges uncut. £2, 2s.

BUTT. Miss Molly. By BEATRICE MAY BUTT. Cheap Edition, 2s.

——— Delicia. By the Author of 'Miss Molly.' Fourth Edition. Crown 8vo, 7s. 6d.

BURTON. The History of Scotland: From Agricola's Invasion to the Extinction of the last Jacobite Insurrection. By JOHN HILL BURTON, Historiographer-Royal for Scotland. New and Enlarged Edition, 8 vols., and Index. Crown 8vo, £3, 3s.

BURTON. History of the British Empire during the Reign of Queen Anne. In 3 vols. 8vo. 36s.

—— The Cairngorm Mountains. Crown 8vo, 3s. 6d.

CAIRD. Sermons. By JOHN CAIRD, D.D., Principal of the University of Glasgow. Fourteenth Thousand. Fcap. 8vo, 5s.

—— Religion in Common Life. A Sermon preached in Crathie Church, October 14, 1855, before Her Majesty the Queen and Prince Albert. Published by Her Majesty's Command. Price One Shilling. Cheap Edition, 3d.

CAMPBELL, Life of Colin, Lord Clyde. *See* General SHADWELL, at page 20.

CARLYLE. Autobiography of the Rev. Dr Alexander Carlyle, Minister of Inveresk. Containing Memorials of the Men and Events of his Time. Edited by JOHN HILL BURTON. 8vo. Third Edition, with Portrait, 14s.

CAUVIN. A Treasury of the English and German Languages. Compiled from the best Authors and Lexicographers in both Languages. Adapted to the Use of Schools, Students, Travellers, and Men of Business; and forming a Companion to all German-English Dictionaries. By JOSEPH CAUVIN, LL.D. & Ph.D., of the University of Göttingen, &c. Crown 8vo, 7s. 6d.

CHARTERIS. Canonicity; or, Early Testimonies to the Existence and Use of the Books of the New Testament. Based on Kirchhoffer's 'Quellensammlung.' Edited by A. H. CHARTERIS, D.D., Professor of Biblical Criticism in the University of Edinburgh. 8vo, 18s.

—— Life of the Rev. James Robertson, D.D., F.R.S.E., Professor of Divinity and Ecclesiastical History in the University of Edinburgh. By Professor CHARTERIS. With Portrait. 8vo, 10s. 6d.

CHETWYND. Life in a German Village. By the Hon. Mrs HENRY WEYLAND CHETWYND, Author of 'Neighbours and Friends,' 'Janie,' 'Mdlle. d'Estanville,' &c. &c. Second Edition. Crown 8vo, 7s. 6d.

CHEVELEY NOVELS, THE.

I. A MODERN MINISTER. 2 vols. bound in cloth, with Twenty-six Illustrations. 17s.

II. SAUL WEIR. 2 vols. bound in cloth. With Twelve Illustrations by F. Barnard. 16s.

CHURCH SERVICE SOCIETY. A Book of Common Order: Being Forms of Worship issued by the Church Service Society. Fourth Edition, 5s.

COLQUHOUN. The Moor and the Loch. Containing Minute Instructions in all Highland Sports, with Wanderings over Crag and Corrie, Flood and Fell. By JOHN COLQUHOUN. Fifth Edition, greatly enlarged. With Illustrations. 2 vols. post 8vo, 26s.

COTTERILL. The Genesis of the Church. By the Right. Rev. HENRY COTTERILL, D.D., Bishop of Edinburgh. Demy 8vo, 16s.

CRANSTOUN. The Elegies of Albius Tibullus. Translated into English Verse, with Life of the Poet, and Illustrative Notes. By JAMES CRANSTOUN, LL.D., Author of a Translation of 'Catullus.' Crown 8vo, 6s. 6d.

—— The Elegies of Sextus Propertius. Translated into English Verse, with Life of the Poet, and Illustrative Notes. Crown 8vo, 7s. 6d.

CRAWFORD. The Doctrine of Holy Scripture respecting the Atonement. By the late THOMAS J. CRAWFORD, D.D., Professor of Divinity in the University of Edinburgh. Third Edition. 8vo, 12s.

—— The Fatherhood of God, Considered in its General and Special Aspects, and particularly in relation to the Atonement, with a Review of Recent Speculations on the Subject. Third Edition, Revised and Enlarged. 8vo, 9s.

CRAWFORD. The Preaching of the Cross, and other Sermons. 8vo, 7s. 6d.

——— The Mysteries of Christianity; being the Baird Lecture for 1874. Crown 8vo, 7s. 6d.

CROSSE. Round about the Carpathians. By ANDREW F. CROSSE, F.C.S. 8vo, with Map of the Author's route, price 12s. 6d.

CUMMING. From Patmos to Paradise; or, Light on the Past, the Present, and the Future. By the Rev. JOHN CUMMING, D.D., F.R.S.E. Crown 8vo, 7s. 6d.

DESCARTES. The Method, Meditations, and Principles of Philosophy of Descartes. Translated from the Original French and Latin. With a New Introductory Essay, Historical and Critical, on the Cartesian Philosophy. By JOHN VEITCH, LL.D., Professor of Logic and Rhetoric in the University of Glasgow. A New Edition, being the Eighth. Price 6s. 6d.

DICKSON. Japan; being a Sketch of the History, Government, and Officers of the Empire. By WALTER DICKSON. 8vo, 15s.

DU CANE. The Odyssey of Homer, Books I.-XII. Translated into English Verse. By Sir CHARLES DU CANE, K.C.M.G. 8vo, 10s. 6d.

EAGLES. Essays. By the Rev. JOHN EAGLES, A.M. Oxon. Originally published in 'Blackwood's Magazine.' Post 8vo, 10s. 6d.

——— The Sketcher. Originally published in 'Blackwood's Magazine.' Post 8vo, 10s. 6d.

ELIOT. Impressions of Theophrastus Such. By GEORGE ELIOT. Fourth Edition. Post 8vo, 10s. 6d. New and cheaper Edition. Crown 8vo, 5s.

——— Adam Bede. Illustrated Edition. 3s. 6d., cloth.

——— The Mill on the Floss. Illustrated Edition. 3s. 6d., cloth.

——— Scenes of Clerical Life. Illustrated Edition. 3s., cloth.

——— Silas Marner: The Weaver of Raveloe. Illustrated Edition. 2s. 6d., cloth.

——— Felix Holt, the Radical. Illustrated Edition. 3s. 6d., cloth.

——— Romola. With Vignette. 3s. 6d., cloth.

——— Middlemarch. Crown 8vo, 7s. 6d.

——— Daniel Deronda. Crown 8vo, 7s. 6d.

——— Works of George Eliot (Cabinet Edition). Complete and Uniform Edition, handsomely printed in a new type, 20 volumes, crown 8vo, price £5. The Volumes are also sold separately, price 5s. each, viz.:—

Romola. 2 vols.—Silas Marner, The Lifted Veil, Brother Jacob. 1 vol.—Adam Bede. 2 vols.—Scenes of Clerical Life. 2 vols.—The Mill on the Floss. 2 vols.—Felix Holt. 2 vols.—Middlemarch. 3 vols.—Daniel Deronda. 3 vols.—The Spanish Gypsy. 1 vol.—Jubal, and other Poems, Old and New. 1 vol.—Theophrastus Such. 1 vol.

——— The Spanish Gypsy. Seventh Edition. Crown 8vo, 7s. 6d., cloth.

——— The Legend of Jubal, and other Poems. New Edition. Fcap. 8vo, 5s., cloth.

——— Wise, Witty, and Tender Sayings, in Prose and Verse. Selected from the Works of GEORGE ELIOT. Fourth Edition, Enlarged. Fcap. 8vo, 6s.

——— The George Eliot Birthday Book. Printed on fine paper, with red border, and handsomely bound in cloth, gilt. Fcap. 8vo, cloth, 3s. 6d. And in French morocco or Russia, 5s.

ESSAYS ON SOCIAL SUBJECTS. Originally published in the 'Saturday Review.' A New Edition. First and Second Series. 2 vols. crown 8vo, 6s. each.

EWALD. The Crown and its Advisers; or, Queen, Ministers, Lords, and Commons. By ALEXANDER CHARLES EWALD, F.S.A. Crown 8vo, 5s.

FERRIER. Philosophical Works of the late James F. Ferrier, A.B. Oxon., Professor of Moral Philosophy and Political Economy, St Andrews. New Edition. Edited by Sir ALEX. GRANT, Bart., D.C.L., and Professor LUSHINGTON. 3 vols. crown 8vo, 34s. 6d.

——— Institutes of Metaphysic. Third Edition. 10s. 6d.

——— Lectures on the Early Greek Philosophy. Second Edition. 10s. 6d.

——— Philosophical Remains, including the Lectures on Early Greek Philosophy. 2 vols., 24s.

FERRIER. George Eliot and Judaism. An Attempt to appreciate 'Daniel Deronda.' By Professor DAVID KAUFMANN, of the Jewish Theological Seminary, Buda-Pesth. Translated from the German by J. W. FERRIER. Second Edition. Crown 8vo, 2s. 6d.

FINLAY. History of Greece under Foreign Domination. By the late GEORGE FINLAY, LL.D., Athens. 6 vols. 8vo—viz.:

Greece under the Romans. B.C. 146 to A.D. 717. A Historical View of the Condition of the Greek Nation from its Conquest by the Romans until the Extinction of the Roman Power in the East. Second Edition, 16s.

History of the Byzantine Empire. A.D. 716 to 1204; and of the Greek Empire of Nicæa and Constantinople, A.D. 1204 to 1453. 2 vols., £1, 7s. 6d.

Greece under Othoman and Venetian Domination. A.D. 1453 to 1821. 10s. 6d.

History of the Greek Revolution of 1830. 2 vols. 8vo, £1, 4s.

FLINT. The Philosophy of History in Europe. Vol. I., containing the History of that Philosophy in France and Germany. By ROBERT FLINT, D.D., LL.D., Professor of Divinity, University of Edinburgh. 8vo, 15s.

——— Theism. Being the Baird Lecture for 1876. Third Edition. Crown 8vo, 7s. 6d.

——— Anti-Theistic Theories. Being the Baird Lecture for 1877. Second Edition. Crown 8vo, 10s. 6d.

FORBES. The Campaign of Garibaldi in the Two Sicilies: A Personal Narrative. By CHARLES STUART FORBES, Commander, R.N. Post 8vo, with Portraits, 12s.

FOREIGN CLASSICS FOR ENGLISH READERS. Edited by Mrs OLIPHANT. Price 2s. 6d.

*Now published*:—I. DANTE. By the Editor.—II. VOLTAIRE. By Major-General Sir E. B. Hamley.—III. PASCAL. By Principal Tulloch.—IV. PETRARCH. By Henry Reeve, C.B.—V. GOETHE. By A. Hayward, Q.C.—VI. MOLIÈRE. By the Editor and F. Tarver, M.A.—VII. MONTAIGNE. By Rev. Canon Collins, M.A.—VIII. RABELAIS. By Walter Besant, M.A.—IX. CALDERON. By E. J. Hasell.—X. SAINT SIMON. By Clifton W. Collins, M.A.—XI. CERVANTES. By the Editor.

*In preparation*:—CORNEILLE AND RACINE. By Henry M. Trollope.—SCHILLER. By James Sime, Author of 'Life of Lessing.'—ROUSSEAU. By Henry Graham.—LA FONTAINE. By Rev. Canon Collins, M.A.—MADAME DE SÉVIGNÉ and MADAME DE STAËL. By Miss Thackeray.

FRASER. Handy Book of Ornamental Conifers, and of Rhododendrons and other American Flowering Shrubs, suitable for the Climate and Soils of Britain. With descriptions of the best kinds, and containing Useful Hints for their successful Cultivation. By HUGH FRASER, Fellow of the Botanical Society of Edinburgh. Crown 8vo, 6s.

GALT. Annals of the Parish. By JOHN GALT. Fcap. 8vo, 2s.

——— The Provost. Fcap. 8vo, 2s.

——— Sir Andrew Wylie. Fcap. 8vo, 2s.

——— The Entail; or, The Laird of Grippy. Fcap. 8vo, 2s.

GARDENER, THE: A Magazine of Horticulture and Floriculture. Edited by DAVID THOMSON, Author of 'The Handy-Book of the Flower-Garden,' &c.; Assisted by a Staff of the best practical Writers. Published Monthly, 6d

GENERAL ASSEMBLY OF THE CHURCH OF SCOTLAND.

——— Family Prayers. Authorised by the General Assembly of the Church of Scotland. A New Edition, crown 8vo, in large type, 4s. 6d. Another Edition, crown 8vo, 2s.

——— Prayers for Social and Family Worship. For the Use of Soldiers, Sailors, Colonists, and Sojourners in India, and other Persons, at home and abroad, who are deprived of the ordinary services of a Christian Ministry. Cheap Edition, 1s. 6d.

——— The Scottish Hymnal. Hymns for Public Worship. Published for Use in Churches by Authority of the General Assembly. Various sizes—viz.: 1. Large type, for pulpit use, cloth, 3s. 6d. 2. Longprimer type, cloth, red edges, 1s. 6d.; French morocco, 2s. 6d.; calf, 6s. 3. Bourgeois type, cloth, red edges, 1s.; French morocco, 2s. 4. Minion type, limp cloth, 6d.; French morocco, 1s. 6d. 5. School Edition, in paper cover, 2d. 6. Children's Hymnal, paper cover 1d. No. 2, bound with the Psalms and Paraphrases, cloth, 3s.; French morocco, 4s. 6d.; calf, 7s. 6d. No. 3, bound with the Psalms and Paraphrases, cloth, 2s.; French morocco, 3s.

——— The Scottish Hymnal, with Music. Selected by the Committees on Hymns and on Psalmody. The harmonies arranged by W. H. Monk. Cloth, 1s. 6d.; French morocco, 3s. 6d. The same in the Tonic Sol-fa Notation, 1s. 6d. and 3s. 6d.

GERARD. Reata: What's in a Name? By E. D. GERARD. Three vols. post 8vo, 25s. 6d.

GLEIG. The Subaltern. By G. R. GLEIG, M.A., late Chaplain-General of her Majesty's Forces. Originally published in 'Blackwood's Magazine.' Library Edition. Revised and Corrected, with a New Preface. Crown 8vo, 7s. 6d.

——— The Great Problem: Can it be Solved? 8vo, 10s. 6d.

GOETHE'S FAUST. Translated into English Verse by Sir THEODORE MARTIN, K.C.B. Second Edition, post 8vo, 6s. Cheap Edition, fcap., 3s. 6d.

——— Poems and Ballads of Goethe. Translated by Professor AYTOUN and Sir THEODORE MARTIN, K.C.B. Third Edition, fcap. 8vo, 6s.

GORDON CUMMING. At Home in Fiji. By C. F. GORDON CUMMING, Author of 'From the Hebrides to the Himalayas.' In 2 vols. post 8vo. With Numerous Illustrations and Map.

GRAHAM. Annals and Correspondence of the Viscount and First and Second Earls of Stair. By JOHN MURRAY GRAHAM. 2 vols. demy 8vo, with Portraits and other Illustrations. £1, 8s.

——— Memoir of Lord Lynedoch. Second Edition, crown 8vo, 5s.

GRANT. Bush-Life in Queensland. By A. C. GRANT. In 2 vols. crown 8vo. [*In the Press.*

GRANT. Incidents in the Sepoy War of 1857-58. Compiled from the Private Journals of the late General Sir HOPE GRANT, G.C.B.; together with some Explanatory Chapters by Captain HENRY KNOLLYS, R.A. Crown 8vo, with Map and Plans, 12s.

GRANT. Memorials of the Castle of Edinburgh. By JAMES GRANT. A New Edition. Crown 8vo, with 12 Engravings, 2s.

HAMERTON. Wenderholme: A Story of Lancashire and Yorkshire Life. By PHILIP GILBERT HAMERTON, Author of 'A Painter's Camp.' A New Edition. Crown 8vo, 6s.

HAMILTON. Lectures on Metaphysics. By Sir WILLIAM HAMILTON, Bart., Professor of Logic and Metaphysics in the University of Edinburgh. Edited by the Rev. H. L. MANSEL, B.D., LL.D., Dean of St Paul's; and JOHN VEITCH, M.A., Professor of Logic and Rhetoric, Glasgow. Sixth Edition. 2 vols. 8vo, 24s.

——— Lectures on Logic. Edited by the SAME. Third Edition. 2 vols. 24s.

——— Discussions on Philosophy and Literature, Education and University Reform. Third Edition, 8vo, 21s.

——— Memoir of Sir William Hamilton, Bart., Professor of Logic and Metaphysics in the University of Edinburgh. By Professor VEITCH of the University of Glasgow. 8vo, with Portrait, 18s.

HAMILTON. Annals of the Peninsular Campaigns. By Captain THOMAS HAMILTON. Edited by F. Hardman. 8vo, 16s. Atlas of Maps to illustrate the Campaigns, 12s.

HAMLEY. The Operations of War Explained and Illustrated. By Sir EDWARD BRUCE HAMLEY, C.B. Fourth Edition, revised throughout. 4to, with numerous Illustrations, 30s.

——— The Story of the Campaign of Sebastopol. Written in the Camp. With Illustrations drawn in Camp by the Author. 8vo, 21s.

——— On Outposts. Second Edition. 8vo, 2s.

——— Wellington's Career; A Military and Political Summary. Crown 8vo, 2s.

——— Lady Lee's Widowhood. Crown 8vo, 2s. 6d.

——— Our Poor Relations. A Philozoic Essay. With Illustrations, chiefly by Ernest Griset. Crown 8vo, cloth gilt, 3s. 6d.

HAMLEY. Guilty, or Not Guilty? A Tale. By Major-General W. G. HAMLEY, late of the Royal Engineers. New Edition. Crown 8vo, 3s. 6d.

——— The House of Lys: One Book of its History. A Tale. Second Edition. 2 vols. crown 8vo. 17s.

HANDY HORSE-BOOK; or, Practical Instructions in Riding, Driving, and the General Care and Management of Horses. By 'MAGENTA.' A New Edition, with 6 Engravings, 4s. 6d.

BY THE SAME.

Our Domesticated Dogs: their Treatment in reference to Food, Diseases, Habits, Punishment, Accomplishments. Crown 8vo, 2s. 6d.

HARBORD. A Glossary of Navigation. Containing the Definitions and Propositions of the Science, Explanation of Terms, and Description of Instruments. By the Rev. J. B. HARBORD, M.A., Assistant Director of Education, Admiralty. Crown 8vo. Illustrated with Diagrams, 6s.

——— Definitions and Diagrams in Astronomy and Navigation. 1s.

——— Short Sermons for Hospitals and Sick Seamen. Fcap. 8vo, cloth, 4s. 6d.

HARDMAN. Scenes and Adventures in Central America. Edited by FREDERICK HARDMAN. Crown 8vo, 6s.

HASTINGS. Poems. By the Lady FLORA HASTINGS. Edited by her SISTER, the late Marchioness of Bute. Second Edition, with a Portrait. Fcap., 7s. 6d.

HAY. The Works of the Right Rev. Dr George Hay, Bishop of Edinburgh. Edited under the Supervision of the Right Rev. Bishop STRAIN. With Memoir and Portrait of the Author. 5 vols. crown 8vo, bound in extra cloth, £1, 1s. Or, sold separately—viz.:

—— The Sincere Christian Instructed in the Faith of Christ from the Written Word. 2 vols., 8s.

—— The Devout Christian Instructed in the Law of Christ from the Written Word. 2 vols., 8s.

—— The Pious Christian Instructed in the Nature and Practice of the Principal Exercises of Piety. 1 vol., 4s.

HEMANS. The Poetical Works of Mrs Hemans. Copyright Editions.

One Volume, royal 8vo, 5s.

The Same, with Illustrations engraved on Steel, bound in cloth, gilt edges, 7s. 6d.

Six Volumes, fcap., 12s. 6d.

Seven Volumes, fcap., with Memoir by her SISTER. 35s.

SELECT POEMS OF MRS HEMANS. Fcap., cloth, gilt edges, 3s.

—— Memoir of Mrs Hemans. By her SISTER. With a Portrait, fcap. 8vo, 5s.

HOLE. A Book about Roses: How to Grow and Show Them. By the Rev. Canon HOLE. With coloured Frontispiece by the Hon. Mrs Francklin. Seventh Edition, revised. Crown 8vo, 7s. 6d.

HOME PRAYERS. By Ministers of the Church of Scotland and Members of the Church Service Society. Fcap. 8vo, price 3s.

HOMER. The Odyssey. Translated into English Verse in the Spenserian Stanza. By PHILIP STANHOPE WORSLEY. Third Edition, 2 vols. fcap., 12s.

—— The Iliad. Translated by P. S. WORSLEY and Professor CONINGTON. 2 vols. crown 8vo, 21s.

HOSACK. Mary Queen of Scots and Her Accusers. Containing a Variety of Documents never before published. By JOHN HOSACK, Barrister-at-Law. A New and Enlarged Edition, with a Photograph from the Bust on the Tomb in Westminster Abbey. 2 vols. 8vo, £1, 1s.

INDEX GEOGRAPHICUS: Being a List, alphabetically arranged, of the Principal Places on the Globe, with the Countries and Subdivisions of the Countries in which they are situated, and their Latitudes and Longitudes. Applicable to all Modern Atlases and Maps. Imperial 8vo, pp. 676, 21s.

JEAN JAMBON. Our Trip to Blunderland; or, Grand Excursion to Blundertown and Back. By JEAN JAMBON. With Sixty Illustrations designed by CHARLES DOYLE, engraved by DALZIEL. Fourth Thousand. Handsomely bound in cloth, gilt edges, 6s. 6d. Cheap Edition, cloth, 3s. 6d. In boards, 2s. 6d.

JOHNSON. The Scots Musical Museum. Consisting of upwards of Six Hundred Songs, with proper Basses for the Pianoforte. Originally published by JAMES JOHNSON; and now accompanied with Copious Notes and Illustrations of the Lyric Poetry and Music of Scotland, by the late WILLIAM STENHOUSE; with additional Notes and Illustrations, by DAVID LAING and C. K. SHARPE. 4 vols. 8vo, Roxburghe binding, £2, 12s. 6d.

JOHNSTON. The Chemistry of Common Life. By Professor J. F. W. JOHNSTON. New Edition, Revised, and brought down to date. By ARTHUR HERBERT CHURCH, M.A. Oxon.; Author of 'Food: its Sources, Constituents, and Uses;' 'The Laboratory Guide for Agricultural Students;' 'Plain Words about Water,' &c. Illustrated with Maps and 102 Engravings on Wood. Complete in One Volume, crown 8vo, pp. 618, 7s. 6d.

——— Professor Johnston's Elements of Agricultural Chemistry and Geology. Eleventh Edition, Revised, and brought down to date. By CHARLES A. CAMERON, M.D., F.R.C.S.I., &c. Fcap. 8vo, 6s. 6d.

——— Catechism of Agricultural Chemistry and Geology. An entirely New Edition, revised and enlarged, by CHARLES A. CAMERON, M.D., F.R.C.S.I., &c. Seventy-eighth Thousand, with numerous Illustrations, 1s.

——— Notes on North America: Agricultural, Economical, and Social. 2 vols. post 8vo, 21s.

KING. The Metamorphoses of Ovid. Translated in English Blank Verse. By HENRY KING, M.A., Fellow of Wadham College, Oxford, and of the Inner Temple, Barrister-at-Law. Crown 8vo, 10s. 6d.

KINGLAKE. History of the Invasion of the Crimea. By A. W. KINGLAKE. Cabinet Edition. Six Volumes, crown 8vo, at 6s. each. The Volumes respectively contain:—

I. THE ORIGIN OF THE WAR between the Czar and the Sultan.
II. RUSSIA MET AND INVADED. With 4 Maps and Plans.
III. THE BATTLE OF THE ALMA. With 14 Maps and Plans.
IV. SEBASTOPOL AT BAY. With 10 Maps and Plans.
V. THE BATTLE OF BALACLAVA. With 10 Maps and Plans.
VI. THE BATTLE OF INKERMAN. With 11 Maps and Plans.

——— History of the Invasion of the Crimea. Vol. VI. Winter Troubles. Demy 8vo, with a Map, 16s.

——— Eothen. A New Edition, uniform with the Cabinet Edition of the 'History of the Crimean War,' price 6s.

KNOLLYS. The Elements of Field-Artillery. Designed for the Use of Infantry and Cavalry Officers. By HENRY KNOLLYS, Captain Royal Artillery; Author of 'From Sedan to Saarbrück,' Editor of 'Incidents in the Sepoy War,' &c. With Engravings. Crown 8vo, 7s. 6d.

LAKEMAN. What I saw in Kaffir-land. By Sir STEPHEN LAKEMAN (MAZHAR PACHA). Post 8vo, 8s. 6d.

LAVERGNE. The Rural Economy of England, Scotland, and Ireland. By LEONCE DE LAVERGNE. Translated from the French. With Notes by a Scottish Farmer. 8vo, 12s.

LEE. Lectures on the History of the Church of Scotland, from the Reformation to the Revolution Settlement. By the late Very Rev. JOHN LEE, D.D., LL.D., Principal of the University of Edinburgh. With Notes and Appendices from the Author's Papers. Edited by the Rev. WILLIAM LEE, D.D. 2 vols. 8vo, 21s.

LEE-HAMILTON. Poems and Transcripts. By EUGENE LEE-HAMILTON. Crown 8vo, 6s.

LEWES. The Physiology of Common Life. By GEORGE H. LEWES, Author of 'Sea-side Studies,' &c. Illustrated with numerous Engravings. 2 vols., 12s.

LOCKHART. Doubles and Quits. By Laurence W. M. Lockhart. With Twelve Illustrations. New Edition. Crown 8vo, 6s.

—— Fair to See: a Novel. Sixth Edition, crown 8vo, 6s.

—— Mine is Thine: a Novel. Sixth Edition, crown 8vo, 6s.

LORIMER. The Institutes of Law: A Treatise of the Principles of Jurisprudence as determined by Nature. By James Lorimer, Regius Professor of Public Law and of the Law of Nature and Nations in the University of Edinburgh. New Edition, revised throughout, and much enlarged. 8vo, 18s.

LYON. History of the Rise and Progress of Freemasonry in Scotland. By David Murray Lyon, Secretary to the Grand Lodge of Scotland. In small quarto. Illustrated with numerous Portraits of Eminent Members of the Craft, and Facsimiles of Ancient Charters and other Curious Documents. £1, 11s. 6d.

M'COMBIE. Cattle and Cattle-Breeders. By William M'Combie, Tillyfour. A New and Cheaper Edition, 2s. 6d., cloth.

MACRAE. A Handbook of Deer-Stalking. By Alexander Macrae, late Forester to Lord Henry Bentinck. With Introduction by Horatio Ross, Esq. Fcap. 8vo, with two Photographs from Life. 3s. 6d.

M'CRIE. Works of the Rev. Thomas M'Crie, D.D. Uniform Edition. Four vols. crown 8vo, 24s.

—— Life of John Knox. Containing Illustrations of the History of the Reformation in Scotland. Crown 8vo, 6s. Another Edition, 3s. 6d.

—— Life of Andrew Melville. Containing Illustrations of the Ecclesiastical and Literary History of Scotland in the Sixteenth and Seventeenth Centuries. Crown 8vo, 6s.

—— History of the Progress and Suppression of the Reformation in Italy in the Sixteenth Century. Crown 8vo, 4s.

—— History of the Progress and Suppression of the Reformation in Spain in the Sixteenth Century. Crown 8vo, 3s. 6d.

—— Sermons, and Review of the 'Tales of My Landlord.' Crown 8vo, 6s.

—— Lectures on the Book of Esther. Fcap. 8vo, 5s.

M'INTOSH. The Book of the Garden. By Charles M'Intosh, formerly Curator of the Royal Gardens of his Majesty the King of the Belgians, and lately of those of his Grace the Duke of Buccleuch, K.G., at Dalkeith Palace. Two large vols. royal 8vo, embellished with 1350 Engravings. £4, 7s. 6d.

Vol. I. On the Formation of Gardens and Construction of Garden Edifices. 776 pages, and 1073 Engravings, £2, 10s.

Vol. II. Practical Gardening. 868 pages, and 279 Engravings, £1, 17s. 6d.

MACKAY. A Manual of Modern Geography; Mathematical, Physical, and Political. By the Rev. Alexander Mackay, LL.D., F.R.G.S. New and Greatly Improved Edition. Crown 8vo, pp. 688. 7s. 6d.

—— Elements of Modern Geography. 46th Thousand, revised to the present time. Crown 8vo, pp. 300, 3s.

—— The Intermediate Geography. Intended as an Intermediate Book between the Author's 'Outlines of Geography,' and 'Elements of Geography.' Sixth Edition, crown 8vo, pp. 224, 2s.

—— Outlines of Modern Geography. 131st Thousand, revised to the Present Time. 18mo, pp. 112, 1s.

—— First Steps in Geography. 69th Thousand. 18mo, pp. 56. Sewed, 4d.; cloth, 6d.

—— Elements of Physiography and Physical Geography. With Express Reference to the Instructions recently issued by the Science and Art Department. 15th Thousand. Crown 8vo, 1s. 6d.

MACKAY. Facts and Dates; or, the Leading Events in Sacred and Profane History, and the Principal Facts in the various Physical Sciences. The Memory being aided throughout by a Simple and Natural Method. For Schools and Private Reference. New Edition, thoroughly Revised. Crown 8vo, 3s. 6d.

MACKENZIE. Studies in Roman Law. With Comparative Views of the Laws of France, England, and Scotland. By LORD MACKENZIE, one of the Judges of the Court of Session in Scotland. Fifth Edition, Edited by JOHN KIRKPATRICK, Esq., M.A. Cantab.; Dr Jur. Heidelb.; LL.B., Edin.; Advocate. 8vo, 12s.

MARMORNE. The Story is told by ADOLPHUS SEGRAVE, the youngest of three Brothers. Third Edition. Crown 8vo, 6s.

MARSHALL. French Home Life. By FREDERIC MARSHALL.

CONTENTS: Servants.—Children.—Furniture.—Food.—Manners.—Language.—Dress.—Marriage. Second Edition. 5s.

MARSHMAN. History of India. From the Earliest Period to the Close of the India Company's Government; with an Epitome of Subsequent Events. By JOHN CLARK MARSHMAN, C.S.I. Abridged from the Author's larger work. Second Edition, revised. Crown 8vo, with Map, 6s. 6d.

MARTIN. Goethe's Faust. Translated by Sir THEODORE MARTIN, K.C.B. Second Edition, crown 8vo, 6s. Cheap Edition, 3s. 6d.

——— Poems and Ballads of Heinrich Heine. Done into English Verse. Printed on *papier vergé*, crown 8vo, 8s.

——— The Odes of Horace. With Life and Notes. Fourth Edition. [*In the press.*

——— Catullus. With Life and Notes. Second Edition, post 8vo, 7s. 6d.

——— The Vita Nuova of Dante. With an Introduction and Notes. Second Edition, crown 8vo, 5s.

——— Aladdin: A Dramatic Poem. By ADAM OEHLENSCHLAEGER. Fcap. 8vo, 5s.

——— Correggio: A Tragedy. By OEHLENSCHLAEGER. With Notes. Fcap. 8vo, 3s.

——— King Rene's Daughter: A Danish Lyrical Drama. By HENRIK HERTZ. Second Edition, fcap., 2s. 6d.

MEIKLEJOHN. An Old Educational Reformer—Dr Bell. By J. M. D. MEIKLEJOHN, M.A., Professor of the Theory, History, and Practice of Education in the University of St Andrews. [*In the press.*

MINTO. A Manual of English Prose Literature, Biographical and Critical: designed mainly to show Characteristics of Style. By W. MINTO, M.A., Professor of Logic in the University of Aberdeen. Second Edition. Crown 8vo, 7s 6d.

——— Characteristics of English Poets, from Chaucer to Shirley. Crown 8vo, 9s.

MITCHELL. Biographies of Eminent Soldiers of the last Four Centuries. By Major-General JOHN MITCHELL, Author of 'Life of Wallenstein.' With a Memoir of the Author. 8vo, 9s.

MOIR. Poetical Works of D. M. MOIR (Delta). With Memoir by THOMAS AIRD, and Portrait. Second Edition, 2 vols. fcap 8vo, 12s.

——— Domestic Verses. New Edition, fcap. 8vo, cloth gilt, 4s. 6d.

——— Lectures on the Poetical Literature of the Past Half-Century. Third Edition, fcap. 8vo, 5s.

——— Life of Mansie Wauch, Tailor in Dalkeith. With 8 Illustrations on Steel, by the late GEORGE CRUIKSHANK. Crown 8vo. 3s. 6d. Another Edition, fcap. 8vo, 1s. 6d.

MONTAGUE. Campaigning in South Africa. Reminiscences of an Officer in 1879 By Captain W. E. MONTAGUE, 94th Regiment, Author of 'Claude Meadowleigh,' &c. 8vo, 10s. 6d.

MONTALEMBERT. Count de Montalembert's History of the Monks of the West. From St Benedict to St Bernard. Translated by Mrs OLIPHANT. 7 vols. 8vo, £3, 17s. 6d.

——— Memoir of Count de Montalembert. A Chapter of Recent French History. By Mrs OLIPHANT, Author of the 'Life of Edward Irving,' &c. 2 vols. crown 8vo. £1, 4s.

MURDOCH. Manual of the Law of Insolvency and Bankruptcy: Comprehending a Summary of the Law of Insolvency, Notour Bankruptcy, Composition-contracts, Trust-deeds, Cessios, and Sequestrations; and the Winding-up of Joint-Stock Companies in Scotland; with Annotations on the various Insolvency and Bankruptcy Statutes; and with Forms of Procedure applicable to these Subjects. By JAMES MURDOCH, Member of the Faculty of Procurators in Glasgow. Fourth Edition, Revised and Enlarged, 8vo, £1.

NEAVES. A Glance at some of the Principles of Comparative Philology. As illustrated in the Latin and Anglican Forms of Speech. By the Hon. Lord NEAVES. Crown 8vo, 1s. 6d.

——— Songs and Verses, Social and Scientific. By an Old Contributor to 'Maga.' Fifth Edition, fcap. 8vo, 4s.

——— The Greek Anthology. Being Vol. XX. of 'Ancient Classics for English Readers.' Crown 8vo. 2s. 6d.

NEW VIRGINIANS, THE. By the Author of 'Estelle Russell,' 'Junia,' &c. In 2 vols., post 8vo.

NICHOLSON. A Manual of Zoology, for the Use of Students. With a General Introduction on the Principles of Zoology. By HENRY ALLEYNE NICHOLSON, M.D., F.R.S.E., F.G.S., &c., Professor of Natural History in the University of St Andrews. Sixth Edition, revised and enlarged. Crown 8vo, pp. 866, with 452 Engravings on Wood, 14s.

——— Text-Book of Zoology, for the Use of Schools. Third Edition, enlarged. Crown 8vo, with 225 Engravings on Wood, 6s.

——— Introductory Text-Book of Zoology, for the Use of Junior Classes. Third Edition, revised and enlarged, with 136 Engravings, 3s.

——— Outlines of Natural History, for Beginners; being Descriptions of a Progressive Series of Zoological Types. Second Edition, with Engravings, 1s. 6d.

NICHOLSON. A Manual of Palæontology, for the Use of Students. With a General Introduction on the Principles of Palæontology. Second Edition. Revised and greatly enlarged. 2 vols. 8vo, with 722 Engravings, £2, 2s.

——— The Ancient Life-History of the Earth. An Outline of the Principles and Leading Facts of Palæontological Science. Crown 8vo, with numerous Engravings, 10s. 6d.

——— On the "Tabulate Corals" of the Palæozoic Period, with Critical Descriptions of Illustrative Species. Illustrated with 15 Lithograph Plates and numerous engravings. Super-royal 8vo, 21s.

NICHOLSON. Redeeming the Time, and other Sermons. By the late MAXWELL NICHOLSON, D.D., Minister of St Stephen's, Edinburgh. Crown 8vo, 7s. 6d.

——— Communion with Heaven, and other Sermons. Crown 8vo, 5s. 6d.

——— Rest in Jesus. Sixth Edition. Fcap. 8vo, 4s. 6d.

OLIPHANT. The Land of Gilead. With Excursions in the Lebanon. By LAURENCE OLIPHANT, Author of 'Lord Elgin's Mission to China and Japan,' &c. With Illustrations and Maps. Demy 8vo, 21s.

——— Piccadilly: A Fragment of Contemporary Biography. With Eight Illustrations by Richard Doyle. Fifth Edition, 4s. 6d. Cheap Edition, in paper cover, 2s. 6d.

——— Russian Shores of the Black Sea in the Autumn of 1852. With a Voyage down the Volga and a Tour through the Country of the Don Cossacks. 8vo, with Map and other Illustrations. Fourth Edition, 14s.

OLIPHANT. Historical Sketches of the Reign of George Second. By Mrs OLIPHANT. Third Edition, 6s.

——— The Story of Valentine; and his Brother. 5s., cloth.

——— Katie Stewart. 2s. 6d.

——— Salem Chapel. 2s. 6d., cloth.

——— The Perpetual Curate. 2s. 6d., cloth.

——— Miss Marjoribanks. 2s. 6d., cloth.

——— The Rector, and the Doctor's Family. 1s. 6d., cloth.

——— John: A Love Story. 2s. 6d., cloth.

OSBORN. Narratives of Voyage and Adventure. By Admiral SHERARD OSBORN, C.B. 3 vols. crown 8vo, 12s. Or separately:—

——— Stray Leaves from an Arctic Journal; or, Eighteen Months in the Polar Regions in Search of Sir John Franklin's Expedition in 1850-51. To which is added the Career, Last Voyage, and Fate of Captain Sir John Franklin. New Edition, crown 8vo, 3s. 6d.

——— The Discovery of a North-West Passage by H.M.S. Investigator, during the years 1850-51-52-53-54. Edited from the Logs and Journals of Captain ROBERT C. M'CLURE. Fourth Edition, crown 8vo, 3s. 6d.

——— Quedah; A Cruise in Japanese Waters: and, The Fight on the Peiho. New Edition, crown 8vo, 5s.

OSSIAN. The Poems of Ossian in the Original Gaelic. With a Literal Translation into English, and a Dissertation on the Authenticity of the Poems. By the Rev. ARCHIBALD CLERK. 2 vols. imperial 8vo, £1, 11s. 6d.

PAGE. Introductory Text-Book of Geology. By DAVID PAGE, LL.D., Professor of Geology in the Durham University of Physical Science, Newcastle. With Engravings on Wood and Glossarial Index. Eleventh Edition, 2s. 6d.

——— Advanced Text-Book of Geology, Descriptive and Industrial. With Engravings, and Glossary of Scientific Terms. Sixth Edition, revised and enlarged, 7s. 6d.

——— Handbook of Geological Terms, Geology, and Physical Geography. Second Edition, enlarged, 7s. 6d.

——— Geology for General Readers. A Series of Popular Sketches in Geology and Palæontology. Third Edition, enlarged, 6s.

——— Chips and Chapters. A Book for Amateurs and Young Geologists. 5s.

——— The Past and Present Life of the Globe. With numerous Illustrations. Crown 8vo, 6s.

——— The Crust of the Earth: A Handy Outline of Geology. Sixth Edition, 1s.

——— Economic Geology; or, Geology in its relation to the Arts and Manufactures. With Engravings, and Coloured Map of the British Islands. Crown 8vo, 7s. 6d.

——— Introductory Text-Book of Physical Geography. With Sketch-Maps and Illustrations. Ninth Edition, 2s. 6d.

——— Advanced Text-Book of Physical Geography. Second Edition. With Engravings. 5s.

PAGET. Paradoxes and Puzzles: Historical, Judicial, and Literary. Now for the first time published in Collected Form. By JOHN PAGET, Barrister-at-Law. 8vo, 12s.

PATON. Spindrift. By Sir J. NOEL PATON. Fcap., cloth, 5s.

——— Poems by a Painter. Fcap., cloth, 5s.

PATTERSON. Essays in History and Art. By R. H. PATTERSON. 8vo, 12s.

PAUL. History of the Royal Company of Archers, the Queen's Body-Guard for Scotland. By JAMES BALFOUR PAUL, Advocate of the Scottish Bar. Crown 4to, with Portraits and other Illustrations. £2, 2s.

PAUL. Analysis and Critical Interpretation of the Hebrew Text of the Book of Genesis. Preceded by a Hebrew Grammar, and Dissertations on the Genuineness of the Pentateuch, and on the Structure of the Hebrew Language. By the Rev. WILLIAM PAUL, A.M. 8vo, 18s.

PERSONALITY. The Beginning and End of Metaphysics, and the Necessary Assumption in all Positive Philosophy. Crown 8vo, 3s.

BY THE SAME.

The Origin of Evil, and Other Sermons. Crown 8vo, 4s. 6d.

PETTIGREW. The Handy-Book of Bees, and their Profitable Management. By A. PETTIGREW. Fourth Edition, Enlarged, with Engravings. Crown 8vo, 3s. 6d.

PHILOSOPHICAL CLASSICS FOR ENGLISH READERS. Companion Series to Ancient and Foreign Classics for English Readers. Edited by WILLIAM KNIGHT, LL.D., Professor of Moral Philosophy, University of St Andrews. In crown 8vo volumes, with portraits, price 3s. 6d. each.

I. DESCARTES. By Professor Mahaffy, Dublin. [*Now ready.*
II. BUTLER. By the Rev. Canon Collins, M.A. [*In January.*

POLLOK. The Course of Time: A Poem. By ROBERT POLLOK, A.M. Small fcap. 8vo, cloth gilt, 2s. 6d. The Cottage Edition, 32mo, sewed, 8d. The Same, cloth, gilt edges, 1s. 6d. Another Edition, with Illustrations by Birket Foster and others, fcap., gilt cloth, 3s. 6d., or with edges gilt, 4s.

PORT ROYAL LOGIC. Translated from the French: with Introduction, Notes, and Appendix. By THOMAS SPENCER BAYNES, LL.D., Professor in the University of St Andrews. Eighth Edition, 12mo, 4s.

POTTS AND DARNELL. Aditus Faciliores: An easy Latin Construing Book, with Complete Vocabulary. By A. W. POTTS, M.A., LL.D., Head-Master of the Fettes College, Edinburgh, and sometime Fellow of St John's College, Cambridge; and the Rev. C. DARNELL, M.A., Head-Master of Cargilfield Preparatory School, Edinburgh, and late Scholar of Pembroke and Downing Colleges, Cambridge. Sixth Edition, fcap. 8vo, 3s. 6d.

——— Aditus Faciliores Graeci. An easy Greek Construing Book, with Complete Vocabulary. Third Edition, fcap. 8vo, 3s.

PRINGLE. The Live-Stock of the Farm. By ROBERT O. PRINGLE. Third Edition, crown 8vo. [*In the press.*

PUBLIC GENERAL STATUTES AFFECTING SCOTLAND, from 1707 to 1847, with Chronological Table and Index. 3 vols. large 8vo, £3, 3s.

PUBLIC GENERAL STATUTES AFFECTING SCOTLAND, COLLECTION OF. Published Annually with General Index.

RAMSAY. Two Lectures on the Genius of Handel, and the Distinctive Character of his Sacred Compositions. Delivered to the Members of the Edinburgh Philosophical Institution. By the Very Rev. DEAN RAMSAY, Author of 'Reminiscences of Scottish Life and Character.' Crown 8vo, 3s. 6d.

RANKINE. A Treatise on the Rights and Burdens Incident to the Ownership of Lands and other Heritages in Scotland. By JOHN RANKINE, M.A., Advocate.' Large 8vo, 40s.

READE. A Woman-Hater. By CHARLES READE. 3 vols. crown 8vo, £1, 5s. 6d. Originally published in 'Blackwood's Magazine.'

REID. A Handy Manual of German Literature. By M. F. REID. For Schools, Civil Service Competitions, and University Local Examinations. Fcap. 8vo, 3s.

RUSTOW. The War for the Rhine Frontier, 1870: Its Political and Military History. By Col. W. RUSTOW. Translated from the German, by JOHN LAYLAND NEEDHAM, Lieutenant R.M. Artillery. 3 vols. 8vo, with Maps and Plans, £1, 11s. 6d.

ST STEPHENS; or, Illustrations of Parliamentary Oratory. A Poem. *Comprising*—Pym—Vane—Strafford—Halifax—Shaftesbury—St John—Sir R. Walpole—Chesterfield—Carteret—Chatham—Pitt—Fox—Burke—Sheridan—Wilberforce—Wyndham—Conway—Castlereagh—William Lamb (Lord Melbourne)—Tierney—Lord Gray—O'Connell—Plunkett—Shiel—Follett—Macaulay—Peel. Second Edition, crown 8vo, 5s.

SANDFORD AND TOWNSEND. The Great Governing Families of England. By J. LANGTON SANDFORD and MEREDITH TOWNSEND. 2 vols. 8vo, 15s., in extra binding, with richly-gilt cover.

SCHETKY. Ninety Years of Work and Play. Sketches from the Public and Private Career of JOHN CHRISTIAN SCHETKY, late Marine Painter in Ordinary to the Queen. By his DAUGHTER. Crown 8vo, 7s. 6d.

SCOTTISH NATURALIST, THE. A Quarterly Magazine of Natural History. Edited by F. BUCHANAN WHITE, M.D., F.L.S. Annual Subscription, free by post, 4s.

SELLAR. Manual of the Education Acts for Scotland. By ALEXANDER CRAIG SELLAR, Advocate. Seventh Edition, greatly enlarged, and revised to the present time. 8vo, 15s.

SELLER AND STEPHENS. Physiology at the Farm; in Aid of Rearing and Feeding the Live Stock. By WILLIAM SELLER, M.D., F.R.S.E., Fellow of the Royal College of Physicians, Edinburgh, formerly Lecturer on Materia Medica and Dietetics; and HENRY STEPHENS, F.R.S.E., Author of 'The Book of the Farm,' &c. Post 8vo, with Engravings, 16s.

SETON. St Kilda: Past and Present. By GEORGE SETON, M.A. Oxon.; Author of the 'Law and Practice of Heraldry in Scotland,' &c. With appropriate Illustrations. Small quarto, 15s.

SHADWELL. The Life of Colin Campbell, Lord Clyde. Illustrated by Extracts from his Diary and Correspondence. By Lieutenant-General SHADWELL, C.B. In 2 vols., 8vo. With Portrait, Map, and Plans. [*In the press.*

SIMPSON. Paris after Waterloo: A Revised Edition of a "Visit to Flanders and the Field of Waterloo." By JAMES SIMPSON, Advocate. With 2 coloured Plans of the Battle. Crown 8vo, 5s.

SMITH. Italian Irrigation: A Report on the Agricultural Canals of Piedmont and Lombardy, addressed to the Hon. the Directors of the East India Company; with an Appendix, containing a Sketch of the Irrigation System of Northern and Central India. By Lieut.-Col. R. BAIRD SMITH, F.G.S., Captain, Bengal Engineers. Second Edition. 2 vols. 8vo, with Atlas in folio, 30s.

SMITH. Thorndale; or, The Conflict of Opinions. By WILLIAM SMITH, Author of 'A Discourse on Ethics,' &c. A New Edition. Crown 8vo, 10s. 6d.

——— Gravenhurst; or, Thoughts on Good and Evil. Second Edition, with Memoir of the Author. Crown 8vo, 8s.

——— A Discourse on Ethics of the School of Paley. 8vo, 4s.

——— Dramas. 1. Sir William Crichton. 2. Athelwold. 3. Guidone. 24mo, boards, 3s.

SOUTHEY. Poetical Works of Caroline Bowles Southey. Fcap. 8vo, 5s.

——— The Birthday, and other Poems. Second Edition, 5s.

——— Chapters on Churchyards. Fcap., 2s. 6d.

SPEKE. What led to the Discovery of the Nile Source. By JOHN HANNING SPEKE, Captain H.M. Indian Army. 8vo, with Maps, &c., 14s.

——— Journal of the Discovery of the Source of the Nile. By J. H. SPEKE, Captain H.M. Indian Army. With a Map of Eastern Equatorial Africa by Captain SPEKE; numerous illustrations, chiefly from Drawings by Captain GRANT; and Portraits, engraved on Steel, of Captains SPEKE and GRANT. 8vo, 21s.

STARFORTH. Villa Residences and Farm Architecture: A Series of Designs. By JOHN STARFORTH, Architect. 102 Engravings. Second Edition, medium 4to, £2, 17s. 6d.

STATISTICAL ACCOUNT OF SCOTLAND. Complete, with Index, 15 vols. 8vo, £16, 16s.
Each County sold separately, with Title, Index, and Map, neatly bound in cloth, forming a very valuable Manual to the Landowner, the Tenant, the Manufacturer, the Naturalist, the Tourist, &c.

STEPHENS. The Book of the Farm; detailing the Labours of the Farmer, Farm-Steward, Ploughman, Shepherd, Hedger, Farm-Labourer, Field-Worker, and Cattleman. By HENRY STEPHENS, F.R.S.E. Illustrated with Portraits of Animals painted from the life; and with 557 Engravings on Wood, representing the principal Field Operations, Implements, and Animals treated of in the Work. A New and Revised Edition, the third, in great part Re-written. 2 vols. large 8vo, £2, 10s.

——— The Book of Farm-Buildings; their Arrangement and Construction. By HENRY STEPHENS, F.R.S.E., Author of 'The Book of the Farm;' and ROBERT SCOTT BURN. Illustrated with 1045 Plates and Engravings. Large 8vo, uniform with 'The Book of the Farm,' &c. £1, 11s. 6d.

——— The Book of Farm Implements and Machines. By J. SLIGHT and R. SCOTT BURN, Engineers. Edited by HENRY STEPHENS. Large 8vo, uniform with 'The Book of the Farm,' £2, 2s.

——— Catechism of Practical Agriculture. With Engravings. 1s.

STEWART. Advice to Purchasers of Horses. By JOHN STEWART, V.S. Author of 'Stable Economy.' 2s. 6d.

——— Stable Economy. A Treatise on the Management of Horses in relation to Stabling, Grooming, Feeding, Watering, and Working. Seventh Edition, fcap. 8vo, 6s. 6d.

STORMONTH. Etymological and Pronouncing Dictionary of the English Language. Including a very Copious Selection of Scientific Terms. For Use in Schools and Colleges, and as a Book of General Reference. By the Rev. JAMES STORMONTH. The Pronunciation carefully Revised by the Rev. P. H. PHELP, M.A. Cantab. Fifth Edition, with enlarged Supplement, containing many words not to be found in any other Dictionary. Crown 8vo, pp. 800. 7s. 6d.

——— The School Etymological Dictionary and Word-Book. Combining the advantages of an ordinary pronouncing School Dictionary and an Etymological Spelling-book. Fcap. 8vo, pp. 254. 2s.

STORY. Graffiti D'Italia. By W. W. STORY, Author of 'Roba di Roma.' Second Edition, fcap. 8vo, 7s. 6d.

——— Nero; A Historical Play. Fcap. 8vo, 6s.

STRICKLAND. Lives of the Queens of Scotland, and English Princesses connected with the Regal Succession of Great Britain. By AGNES STRICKLAND. With Portraits and Historical Vignettes. 8 vols. post 8vo, £4, 4s.

STURGIS. John-a-Dreams. A Tale. By JULIAN STURGIS. New Edition, crown 8vo, 3s. 6d.

——— An Accomplished Gentleman. Second Edition. Post 8vo, 7s. 6d.

——— Little Comedies. Crown 8vo, 5s.

SUTHERLAND. Handbook of Hardy Herbaceous and Alpine Flowers, for general Garden Decoration. Containing Descriptions, in Plain Language, of upwards of 1000 Species of Ornamental Hardy Perennial and Alpine Plants, adapted to all classes of Flower-Gardens, Rockwork, and Waters; along with Concise and Plain Instructions for their Propagation and Culture. By WILLIAM SUTHERLAND, Gardener to the Earl of Minto; formerly Manager of the Herbaceous Department at Kew. Crown 8vo, 7s. 6d.

SWAINSON. A Handbook of Weather Folk-Lore. Being a Collection of Proverbial Sayings in various Languages relating to the Weather, with Explanatory and Illustrative Notes. By the Rev. C. SWAINSON, M.A., Vicar of High Hurst Wood. Fcap. 8vo, Roxburghe binding, 6s. 6d.

SWAYNE. Lake Victoria: A Narrative of Explorations in Search of the Source of the Nile. Compiled from the Memoirs of Captains Speke and Grant. By GEORGE C. SWAYNE, M.A., late Fellow of Corpus Christi College, Oxford. Illustrated with Woodcuts and Map. Crown 8vo, 7s. 6d.

TAYLOR. Destruction and Reconstruction: Personal Experiences of the Late War in the United States. By RICHARD TAYLOR, Lieutenant-General in the Confederate Army. 8vo, 10s. 6d.

TAYLOR. The Story of My Life. By the late Colonel MEADOWS TAYLOR, Author of 'The Confessions of a Thug,' &c. &c. Edited by his Daughter. Third Edition, post 8vo, 9s.

THOLUCK. Hours of Christian Devotion. Translated from the German of A. Tholuck, D.D., Professor of Theology in the University of Halle. By the Rev. ROBERT MENZIES, D.D. With a Preface written for this Translation by the Author. Second Edition, crown 8vo, 7s. 6d.

THOMSON. Handy-Book of the Flower-Garden: being Practical Directions for the Propagation, Culture, and Arrangement of Plants in Flower-Gardens all the year round. Embracing all classes of Gardens, from the largest to the smallest. With Engraved and Coloured Plans, illustrative of the various systems of Grouping in Beds and Borders. By DAVID THOMSON, Gardener to his Grace the Duke of Buccleuch, K.G., at Drumlanrig. Third Edition, crown 8vo, 7s. 6d.

——— The Handy-Book of Fruit-Culture under Glass: being a series of Elaborate Practical Treatises on the Cultivation and Forcing of Pines, Vines, Peaches, Figs, Melons, Strawberries, and Cucumbers. With Engravings of Hothouses, &c., most suitable for the Cultivation and Forcing of these Fruits. Crown 8vo, with Engravings, 7s. 6d.

THOMSON. A Practical Treatise on the Cultivation of the Grape-Vine. By WILLIAM THOMSON, Tweed Vineyards. Ninth Edition, 8vo, 5s.

TOM CRINGLE'S LOG. A New Edition, with Illustrations. Crown 8vo, 6s. Cheap Edition, 2s.

TRANSACTIONS OF THE HIGHLAND AND AGRICULTURAL SOCIETY OF SCOTLAND. Published annually, price 5s.

TULLOCH. Rational Theology and Christian Philosophy in England in the Seventeenth Century. By JOHN TULLOCH, D.D., Principal of St Mary's College in the University of St Andrews; and one of her Majesty's Chaplains in Ordinary in Scotland. Second Edition. 2 vols. 8vo, 28s.

——— Some Facts of Religion and of Life. Sermons Preached before her Majesty the Queen in Scotland, 1866-76. Second Edition, crown 8vo, 7s. 6d.

TULLOCH. The Christian Doctrine of Sin; being the Croall Lecture for 1876. Crown 8vo, 6s.

—— Theism. The Witness of Reason and Nature to an All-Wise and Beneficent Creator. 8vo, 10s. 6d.

TYTLER. The Wonder-Seeker; or, The History of Charles Douglas. By M. FRASER TYTLER, Author of 'Tales of the Great and Brave,' &c. A New Edition. Fcap., 3s. 6d.

VIRGIL. The Æneid of Virgil. Translated in English Blank Verse by G. K. RICKARDS, M.A., and Lord RAVENSWORTH. 2 vols. fcap. 8vo, 10s.

WALFORD. Mr Smith: A Part of his Life. By L. B. WALFORD. Cheap Edition, 3s. 6d.

—— Pauline. Fifth Edition. Crown 8vo, 6s.

—— Cousins. Cheaper Edition. Crown 8vo, 6s.

—— Troublesome Daughters. Cheaper Edition. Crown 8vo, 6s.

WARREN'S (SAMUEL) WORKS. People's Edition, 4 vols. crown 8vo, cloth, 18s. Or separately:—

Diary of a Late Physician. 3s. 6d. Illustrated, crown 8vo, 7s. 6d.

Ten Thousand A-Year. 5s.

Now and Then. The Lily and the Bee. Intellectual and Moral Development of the Present Age. 4s. 6d.

Essays: Critical, Imaginative, and Juridical. 5s.

WARREN. The Five Books of the Psalms. With Marginal Notes. By Rev. SAMUEL L. WARREN, Rector of Esher, Surrey; late Fellow, Dean, and Divinity Lecturer, Wadham College, Oxford. Crown 8vo, 5s.

WELLINGTON. Wellington Prize Essays on "the System of Field Manœuvres best adapted for enabling our Troops to meet a Continental Army." Edited by Sir EDWARD BRUCE HAMLEY, C.B. 8vo, 12s. 6d.

WESTMINSTER ASSEMBLY. Minutes of the Westminster Assembly, while engaged in preparing their Directory for Church Government, Confession of Faith, and Catechisms (November 1644 to March 1649). Printed from Transcripts of the Originals procured by the General Assembly of the Church of Scotland. Edited by the Rev. ALEX. T. MITCHELL, D.D., Professor of Ecclesiastical History in the University of St Andrews, and the Rev. JOHN STRUTHERS, LL.D., Minister of Prestonpans. With a Historical and Critical Introduction by Professor Mitchell. 8vo, 15s.

WHITE. The Eighteen Christian Centuries. By the Rev. JAMES WHITE, Author of 'The History of France.' Seventh Edition, post 8vo, with Index, 6s.

—— History of France, from the Earliest Times. Sixth Thousand, post 8vo, with Index, 6s.

WHITE. Archæological Sketches in Scotland—Kintyre and Knapdale. By Captain T. P. WHITE, R.E., of the Ordnance Survey. With numerous Illustrations. 2 vols. folio, £4, 4s. Vol. I., Kintyre, sold separately, £2, 2s.

WILLS AND GREENE. Drawing-room Dramas for Children. By W. G. WILLS and the Hon. Mrs GREENE. Crown 8vo, 6s.

WILSON. The "Ever-Victorious Army:" A History of the Chinese Campaign under Lieut.-Col. C. G. Gordon, and of the Suppression of the Tai-ping Rebellion. By ANDREW WILSON, F.A.S.L. 8vo, with Maps, 15s.

—— The Abode of Snow: Observations on a Journey from Chinese Tibet to the Indian Caucasus, through the Upper Valleys of the Himalaya. New Edition. Crown 8vo, with Map, 10s. 6d.

WILSON. Works of Professor Wilson. Edited by his Son-in-Law, Professor FERRIER. 12 vols. crown 8vo, £2, 8s.

—— Christopher in his Sporting-Jacket. 2 vols., 8s.

—— Isle of Palms, City of the Plague, and other Poems. 4s.

—— Lights and Shadows of Scottish Life, and other Tales. 4s.

—— Essays, Critical and Imaginative. 4 vols., 16s.

—— The Noctes Ambrosianæ. Complete, 4 vols., 14s.

—— The Comedy of the Noctes Ambrosianæ. By CHRISTOPHER NORTH. Edited by JOHN SKELTON, Advocate. With a Portrait of Professor Wilson and of the Ettrick Shepherd, engraved on Steel. Crown 8vo, 7s. 6d.

—— Homer and his Translators, and the Greek Drama. Crown 8vo, 4s.

WINGATE. Annie Weir, and other Poems. By DAVID WINGATE. Fcap. 8vo, 5s.

—— Lily Neil. A Poem. Crown 8vo, 4s. 6d.

WORSLEY. Poems and Translations. By PHILIP STANHOPE WORSLEY, M.A. Edited by EDWARD WORSLEY. Second Edition, enlarged. Fcap. 8vo, 6s.

WYLDE. A Dreamer. By KATHARINE WYLDE. In 3 vols., post 8vo, 25s. 6d.

YOUNG. Songs of Béranger done into English Verse. By WILLIAM YOUNG. New Edition, revised. Fcap. 8vo, 4s. 6d.

YULE. Fortification: for the Use of Officers in the Army, and Readers of Military History. By Col. YULE, Bengal Engineers. 8vo, with numerous Illustrations, 10s. 6d.

Printed in Dunstable, United Kingdom

91218925R00181